BIOTECHNOLOGY IN AGRICULTURE, INDUSTRY AND MEDICINE

NATURAL ENEMIES

IDENTIFICATION, PROTECTION STRATEGIES AND ECOLOGICAL IMPACTS

Biotechnology in Agriculture, Industry and Medicine

Additional books in this series can be found on Nova's website under the Series tab.

Additional e-books in this series can be found on Nova's website under the eBook tab.

BIOTECHNOLOGY IN AGRICULTURE, INDUSTRY AND MEDICINE

NATURAL ENEMIES

IDENTIFICATION, PROTECTION STRATEGIES AND ECOLOGICAL IMPACTS

SÓNIA A. P. SANTOS
EDITOR

nova publishers
New York

Library of Congress Cataloging-in-Publication Data

Names: Santos, Sónia A. P.
Title: Natural enemies : identification, protection strategies and ecological impacts / editor, Sónia A.P. Santos (CIMO/School of Agriculture, Polytechnic Institute of Bragança, Bragança, Portugal).
Description: Hauppauge, New York : Nova Science Publisher's, Inc., [2016] | Series: Biotechnology in agriculture, industry and medicine | Includes bibliographical references and index.
Identifiers: LCCN 2016034018 (print) | LCCN 2016046325 (ebook) | ISBN 9781634859219 (hardcover) | ISBN 9781634859448 (ebook) | ISBN 9781634859448
Subjects: LCSH: Biodiversity. | Insect-plant relationships.
Classification: LCC QH541.15.B56 N39 2016 (print) | LCC QH541.15.B56 (ebook) | DDC 577--dc23
LC record available at https://lccn.loc.gov/2016034018

Published by Nova Science Publishers, Inc. † New York

CONTENTS

PREFACE

Natural enemies are organisms that kill or reduce the activity of other organisms and are particularly useful and functionally relevant in agricultural ecosystems due to their contribution to biological control. Every year, arthropod pest outbreaks cause serious losses in agricultural production and to the income of farmers. Sustainable crop protection strategies, through the conservation or enhancement of natural enemies in agroecosystems, can increase production and food quality and security.

The community of natural enemies that inhabits agroecosystems is usually complex because it includes quite diverse taxonomic groups of pathogens, predators and parasitoids, living in structured assemblages that include dominant, accessory and vestigial species. This complexity can be mainly due to multiple trophic interactions established among natural enemies and their prey/hosts. Predators feed on several prey throughout their lifespan and can be more generalists, feeding on a broad range of prey, or specialists, with a demarcated preference for one type of prey, while a parasitoid specimen kills one host during its developmental stages, but it usually is more species-specific than predators. Trophic interactions influence biological control since natural enemies can show preferences for certain food items other than pests. Moreover, the efficiency of natural enemies can depend on their recruitment into the crop and maintenance of their populations at an adequate level. Most of them require non-prey/non-host food resources such as pollen and nectar from flowers and honeydew secreted by phytophagous hemipterans as well as shelters and overwintering places. Several studies demonstrated that spontaneous herbaceous plants occurring as ground cover plants and landscape patches surrounding the arable area can positively influence the abundance of natural enemies in the crop and, consequently, biological control of pests.

When the habitat is disturbed by periodical cuts, tillage or pesticide application, arthropods will be affected and, if they survive, they may need to move to neighboring areas. Pesticides are also deleterious factors for natural enemies and influence behavior, reproduction parameters and, ultimately, survival.

This book includes a broad range of ecological issues about the identification, protection strategies and impacts of agricultural practices on important groups of natural enemies. The core focus includes specific studies on predatory species (e.g., mites, spiders, anthocorids, nabids, carabids, and coccinellids) as well as parasitoid species (hymenopterans). Chapter 1 presents a case study performed in Spanish olive groves with new approaches towards the promotion of biological control mechanisms through the conservation and restoration of natural ecological infrastructures. Chapter 2 focus on the biological control of *Oligonychus perseae* (Tuttle, Baker and Abbatiello), a herbivore mite specific to avocados, performed by predatory mite species (*Euseius stipulatus* (Athias-Henriot), *E. scutalis* (Athias-Henriot), *Neoseiulus californicus* (McGregor), *Galendromus annectens* (De Leon) and *G. helveolus* (Chant)) in a climate change scenario. Chapter 3 provides an overview of the effects of the most common agricultural practices on the biodiversity and life-history parameters of spiders, one of the most abundant and diverse groups of predators in agroecosystems. Chapter 4 shows that crop management can influence the movement of natural enemies that can migrate to surrounding areas. Methods used to determine the movement of natural enemies, their characteristics, as well as advantages and disadvantages of each method are described. Chapter 5 discusses several management strategies required to increase the abundance of the parasitoid species, *Psyttalia concolor* (Szépligeti) (Hymenoptera: Braconidae), as potential biological control agent of the key pest of the olive tree, the olive fruit fly *Bactrocera oleae* (Rossi) (Diptera: Tephritidae). Chapter 6 reviews the communities of native parasitoid associated with *Dryocosmus kuriphilus* Yasumatsu (Hymenoptera: Cynipidae), a gall-inducing insect, native to China, that attacks buds of trees from the genus *Castanea* in several invaded countries of Asia, Europe and the USA.

Last but not least, I would like to acknowledge the generous contributions of all authors who turned possible the edition of this book.

In: Natural Enemies
Editor: Sónia A.P. Santos

ISBN: 978-1-63485-921-9

Chapter 1

POTENTIAL OF ECOLOGICAL INFRASTRUCTURES TO RESTORE CONSERVATION BIOLOGICAL CONTROL: CASE STUDY IN SPANISH OLIVE GROVES

Daniel Paredes[1,*], Luis Cayuela[2] and Mercedes Campos[1]

[1]Departamento de Protección Ambiental,
Estación Experimental del Zaidín, CSIC, Granada, Spain
[2]Área de Biodiversidad y Conservación,
Universidad Rey Juan Carlos I, Móstoles, Spain

ABSTRACT

World population is exponentially growing and with it, the need for quality food that supports this expansion. Current agricultural techniques rely on a massive use of chemical inputs that at the end jeopardise the sustainability of crops. New approaches are oriented towards the promotion of natural control mechanisms that guarantee biodiversity through the conservation and restoration of natural habitats. In this sense, conservation biological control is the most promising option to meet an equilibrium between production and biodiversity conservation. It is based on the creation of a suitable environment to balance the relationship

* Corresponding author: Departamento de Protección Ambiental. Estación Experimental del Zaidín, CSIC, Profesor Albareda nº 1, 18008, Granada. Spain. Email: daniel.paredes@eez.csic.es.

between crop, pests and natural enemies to avoid fluctuations that can bring about production losses. In this chapter we review the state of the art of this strategy and we present a case study in Andalusian olive groves. Along the document we present the controversies that may arise after the implementation of this strategy and we discuss about possible solutions.

Keywords: ecosystem services, biodiversity, ground cover, landscape, natural enemies, crop protection

1. Introduction

The use of synthetic chemical products to control pests and diseases has led to the consolidation of farming methods based on a wide use of these inputs. Along with this practice, a number of problems that threaten both sustainability and quality of the crops and the health of people and natural systems have appeared (Meehan et al., 2011). To stop this trend, scientific and political professionals are demanding a change to pest management techniques orientated towards more sustainable strategies. Among the possible solutions, biological control is being studied and, between all the possible strategies of this method, conservation biological control is the most promising alternative to reduce the use of chemical products.

Conservation biological control (CBC) differs from other biological control strategies in that it is not based on an insect release, but rather to establish, by applying certain methods, an adequate environment where to develop the crop (Barbosa, 1998). The application of these methods does not exert its own action to control pests but promotes abundance, diversity and effectiveness of natural enemies already present in the agricultural ecosystem for optimal performance against pests (Bianchi et al., 2006; Landis et al., 2000). An increase in plant diversity in the agroecosystem would result in increasing the resources that natural enemies can exploit and, consequently, in improving natural pest control. In traditional farming systems, biodiversity has been used to protect crops from pests and diseases, minimizing the risk of crop loss, producing a varied diet and diversifying sources of income (Altieri, 1991). Therefore, in order to perform a conservation biological control strategy, it is not necessary to carry out specific actions but create an environment to balance the relationship between crop, pests and natural enemies to avoid fluctuations that can bring about production losses (Paredes et al., 2013a). Landis et al. (2000), suggested several resources that have to be enhanced to promote conservation biological control: shelter and nesting sites, supplementary food (nectar, pollen) and alternative hosts or prey. The best way to provide these resources is through the installation and management of ecological infrastructures. These are natural or seminatural non-crop lands that conserve ecosystem functions and provide associated services.

Perennial crops, which are subject to lower levels of disturbance than annual crops, are potentially more suitable for carrying out a conservation biological control strategy. The most commonly used approach has been the installation and management of a ground cover. This normally can be in the form of a strip between tree lines or, less commonly, continuous on the ground (Smith et al., 1996). They are mainly composed of planted species belonging to the families Poaceae and Fabaceae although there are studies in which a natural and spontaneous cover was allowed to grow (Paredes et al., 2013b; Rieux et al., 1999; Silva et al., 2010). Other studies chose to design ground covers with aromatic plants (Song et al., 2010). The study of ground covers has produced very positive results in the abundance and diversity of natural enemies in pear (Rieux et al., 1999; Song et al., 2010), pecan (Smith et al., 1996), apple (Bone et al., 2009), vineyards (Danne et al., 2010), citrus (Silva et al., 2010) and olive crops (Paredes et al., 2013b). However, despite any increase in the abundance of predatory and parasitic insects led by the use of ground covers, it is still uncertain whether this will translate into reduced pest densities. Much of the existing literature focuses on the effects of ground covers on natural enemies, implicitly assuming that an increase in their abundance or diversity will necessarily lead to a decrease in pest abundance. Those studies that explicitly investigate the effects of ground cover on herbivore abundance provide inconsistent evidence (Paredes et al., 2015a). There are two reasons that could explain these controversial results: the influence of the landscape context and the relationship between biodiversity and biological control.

If the crop is located within a very diverse landscape (heterogeneous), the chances of a successful conservation biological control will be greater than those that would exist if the agricultural activity has developed within a homogenous landscape. Thies and Tscharntke (1999) provided the first evidence of this fact and determined that pest damage was negatively related to the landscape diversity that was surrounding the crop. Other studies in the same subject, have also found positive results from the abundance of natural enemies and diversity on the landscape scale, as shown by the latest reviews published in this area (Bianchi et al., 2006; Chaplin-Kramer et al., 2011; Rusch et al., 2016).

The fact that the landscape context has an effect on the quality of conservation biological control has usually been unnoticed in studies that implemented ground covers. Most of the studies have focused on non-crop vegetation (landscape approach) or on the effect that ground cover installation and management can have on natural enemy performance (farm approach) (Paredes et al., 2013b; Woltz et al., 2012). Many of the studies that have assessed conservation biological control techniques have not taken into account the landscape context, thus ignoring its potential (both positive and negative) synergistic effects between seminatural habitats and the implemented techniques (Altieri et al., 2005). However, recent works on the subject already have this possibility in annual crop (Woltz et al., 2012) and perennial crops (Paredes et al., 2013b; Paredes et al., 2013c).

According to the theory, an increase in the diversity of natural enemies will have consequences for biological pest control because increasing the number of species in the community of natural enemies will increase the chances that some of these species perform effective pest control (Loureau, 1998; Straub et al., 2008). However, in most cases, the relationship between biodiversity and biological control is not usually successful (Moonen & Bàrberi, 2008) and, in cases where the biological control is effective, there is one natural enemy or a small group of them that are exerting such control (Denoth et al., 2002; Paredes et al., 2015b). Different ecosystem processes can underlie the relationship between biodiversity making it positive, neutral or even negative.

A key factor that can lead to a negative relationship between biodiversity and biological control is intraguild predation. Intraguild predation occurs when a predator consumes another predator, which competes for the same prey (Polis & Holt, 1992). This fact distorts the first predator′s action to the target species (pest), and decreases the action of the second predator on the pest by the decline of its abundance. Thus, a biodiversity increase in the agroecosystem will increase the chances that this process occurs, reducing the chances of an effective biological control. However, an increase in biodiversity would also increase the possibility that a functional redundancy process would occur. This happens when more than one species have the same functional role within the agroecosystem (Walker, 1992). With this situation, an increase in natural enemy diversity will not transfer into a decrease in pest abundance because the ways to reduce it are overlapped, thus leading to a neutral relationship between biodiversity and biological control. It is also conceivable that an increase in diversity would increase the range of functional strategies of consumption of the pest by different natural enemies. For example, a natural enemy can chase prey while another can ambush, lurking and waiting. Thus, a niche complementarity process would lead to a positive relationship between biodiversity and biological control (Straub et al., 2008).

2. Ground Cover and Conservation Biological Control in Olive Groves

To assess the ability of ecological infrastructures for conservation biological control of pests is not an easy task, as there are a multitude of processes, both direct and indirect, that may influence the effect of these structures on populations of different groups of arthropods. However, there are new studies where the implementation of these techniques has great potential but has been little studied. These new works show an innovative approach in several aspects. First, and to the extent of our knowledge, this is one of the first works in which the combined effects of different ecological infrastructures on arthropod abundance are analysed. Second, the detection of controversies (increased abundance of natural enemies in areas with ground cover not resulting in a decrease in the abundance of pests in the same areas) has led to the study of the relationships between the natural enemy community components and those with pests. Finally, a broad timescale (six years of observations) and spatial (over 2,528 olive trees scattered throughout the Autonomous Community of Andalusia) is used to analyse the effect of ground cover on different pest species, which has contributed to an overview of the role of this structure in conservation biological control.

The olive tree is a crop of particular importance in the Mediterranean region and especially in Spain, a country that is the top world producer with 6,538 tons on average between 2010 and 2014, more than a third of the world production (34.2%) (FAOSTAT, 2014). It also offers optimal characteristics for the implementation of a conservation biological control strategy based on habitat management. Being a perennial crop, the installation and management of a ground cover in the middle of the tree rows is easy. In addition, olive groves are characterized by an abundant and diverse community of natural enemies with a parasitic complex represented by 300-400 species of Hymenoptera order and a predator complex consisting of different orders (Arambourg, 1986). Among the main groups of predators we can find spiders, which are the most diverse with 217 species, followed by Coleoptera order with 30 species, the group of ants with 24, Hemiptera with 11 and Neuroptera with 13 species (Cárdenas, 2008; Cotes et al., 2010; Morris et al., 1999; Redolfi et al., 1999; Rei et al., 2010; Ruiz-Torres & Montiel-Bueno, 2000; Santos et al., 2007). However, several insect species attack and damage olive trees, with the key pest species being the olive fly *Bactrocera oleae* (Rossi) (Diptera: Tephritidae), the olive moth *Prays oleae* Bern. (Lepidoptera: Praydidae), and the olive scale *Saissetia oleae* Olivier (Hemiptera: Coccidae). These three phytophagous insects are widely distributed and usually cause high crop damage by reducing the amount and quality of production (Arambourg, 1986; Tzanakakis, 2006). In addition, there are other secondary pests that can cause important losses more sporadically. Among them are species such as the olive psyllid *Euphyllura olivina* Costa (Hemiptera: Psyllidae), the olive borer *Phloeotribus scarabaeoides* Bern. (Coleoptera: Scolytidae) or *Euzophera pinguis* Haw. (Lepidoptera: Pyralidae) (Alvarado et al., 2008; Campos, 2011; Civantos, 1999).

The olive grove presents varied typologies of landscapes (Duarte et al., 2009). If we consider that greater structural diversity (trees, spontaneous ground cover, patches of natural vegetation, etc.) allows a greater diversity of habitat and specific benefits, it is reasonable to think that each type of olive grove shall raise its particular way of relating to biodiversity. However, the benefits of landscape complexity have been poorly quantified in the olive grove so it is necessary to carry out specific studies and determine the weight of each of these ecological infrastructures in relation to conservation biological control, which will develop targeted measures to enhance the action of the auxiliary fauna and therefore the sustainability of the olive grove (Duarte et al., 2009).

Synergies between ground cover and adjacent natural vegetation patches have been detected in some groups of natural enemies. Spiders and parasitoids were more abundant in plots with ground cover while this structure did not affect Heteroptera predators. Large patches of woody and herbaceous vegetation patches influenced natural enemy abundance more than small patches composed of shrubs. The effect of adjacent natural vegetation was different for each group and relied on the presence of a ground cover. When both structures were present in the crop, parasitoids and spiders were positively influenced by the surrounding vegetation, while this effect was smaller or even negative in areas of bare soil, especially in the case of herbaceous vegetation patches for spiders (Paredes et al., 2013a).

Considering the effects of ground cover on spiders and parasitoids, one would expect that in areas where this structure was present, pest populations were lower than in areas where the ground was bare. However, in olive groves, ground cover did not display an effect on the abundance of *P. oleae* and *E. olivina* (Aldebis et al., 2004; Rodríguez et al., 2009; Paredes et al., 2013c). This could reflect a lack of synchronisms between pest and natural enemies that have shown a response to the presence of ground cover (Perdikis et al., 2011).

However, natural vegetation patches displayed a positive effect on biological pest control in olive groves. In particular, large areas of woody vegetation reduced the abundance of *P. oleae* and small patches of woody vegetation decreased the abundance of *E. olivina*. The effect of other ecological infrastructures showed variability between years in the abundance of the two pest species which may indicate that weather conditions play an important role in modulating the response of pests to their environment (Logan et al., 2006; Martinat, 1987).

The relationship between different natural enemy groups and pests has been explored in order to determine the source of controversy. *E. olivina* was associated with the predatory species *Anthocoris nemoralis* Fabricius (Heteroptera: Anthocoridae), as occurred in other crops such as pear (Agustí et al., 2003; Shaltiel & Coll, 2004). *A. nemoralis* overwinters in the adult stage in the vegetation adjacent to crops (Horton & Lewis, 2000; Paredes et al., 2013b). Probably, its presence in the crop is mediated by the release of herbivore-induced plant volatiles (HIPV's) (Drukker et al., 1995). Consequently, when *E. olivina* reaches high abundance on the tree, the injured tissues can emit a blend of HIPV's that attracts *A. nemoralis*. This is of great importance since *A. nemoralis* has also been related to low abundances of *P. oleae*. When adults of *A. nemoralis* appear on the tree at the beginning of the season, they lay the eggs of the next generation, and because of their small size, new-borns have no available prey except *P. oleae* eggs, thus reducing the abundance of this pest. As a result of this process, *A. nemoralis* could become a very effective biological control agent against the anthophagous generation of *P. oleae* (Morris et al., 1999; Paredes et al., 2015b). In addition to *A. nemoralis*, another group of natural enemies has been identified in a set of predators consisting of two families of spiders, Araneidae and Linyphiidae, and the lacewing *Chrysoperla carnea* Stephen (Neuroptera: Chrysopidae) that were also associated with a low abundance of *P. oleae*. These groups could play complementary roles determined by the way they capture their prey

(Straub et al., 2008). Nevertheless, further information is needed to confirm the trends observed in these studies.

Apparently, olive pest abundance is more affected by the presence of patches of natural vegetation and the natural enemies that they can harbour than by the presence of a ground cover within the crop. However, there are certain circumstances (landscape and climate) that could determine the role of ground cover on biological pest control. These could be masking the true effect of this structure, drawing on non-valid conclusions if studies do not take into account these characteristics or are developed at a local scale. The analysis on the effect of ground cover in a large study area and a large time scale provided much more robust evidence about the usefulness of ground cover for the conservation biological control of olive pests. Results indicated that the ground cover, as a single factor and compared with bare soil, was not effective in reducing pest abundance in olive groves (Paredes et al., 2015a). The variability explained by random factors related to the specific conditions of the plot (local scale), the landscape context (landscape scale) and climatic variability between years (regional) was much higher than those explained by ground cover.

Locally, the physical characteristics of the land (topography, microclimatic characteristics, soil type, cultivar), as well as the farm management (intensity of pesticides applied, history of land use), can have a significant effect on pest abundance. Ground cover composition (Drukker et al., 1995), density (Colloff et al., 2013), width of the strip (Hardman et al., 2011), or height of the herbs can also have an effect on the abundance of pests (Tsitsilas et al., 2011). Another factor that can influence pest populations is the historical application of pesticides (Croft & Brown, 1975). An abuse of these substances in the recent past could reduce populations of natural enemies to levels ineffective for biological control. In addition, the lack of natural vegetation patches can increase the injurious effect of pesticides because natural enemies do not have places to shelter when olive groves are sprayed (Landis et al., 2000; Bianchi et al., 2006; Simon et al., 2010).

On a landscape scale, structures are able to influence the colonization of natural enemies into the crop (Bianchi et al., 2006; Perovic et al., 2010; Thies & Tscharntke, 1999). Differences in the landscape context can have great influence on the abundance of pests (Bianchi et al., 2006; Rusch et al., 2013; Thies & Tscharntke, 1999). Natural vegetation patches characteristics; such as edge, density and landscape diversity can help to reduce the abundance of *B. oleae* (Boccaccio & Petacchi, 2009; Ortega & Pascual, 2014). This does not necessarily mean that the higher the density of natural vegetation the greater biological control because sometimes it may be the opposite effect due to negative interactions between organisms belonging to agroecosystem (Martin et al., 2013) or that these habitats have the opposite effect by removing, instead of providing natural enemies in the crop (Karp et al., 2013). Resolving these complexities requires analysing the effects of natural habitat adjacent to crops on predators, pests and predation rates (Bianchi et al., 2006; Chaplin-Kramer et al., 2011; Karp et al., 2013).

Conclusion

Although a clear effect of ground cover on biological control of olive pests was not observed, it can help to increase local habitat heterogeneity, showing its usefulness to maintain populations of beneficial arthropods within the olive grove. A more detailed analysis of its composition and management in order to optimize its effect on conservation biological control would be advisable because it would draw on conclusions capable of generating useful management guidelines for stakeholders. However, ground cover is recommended to enhance other ecosystem services such as nutrient recycling, prevention of erosion (Culler et al., 2008; Hartwig & Ammon, 2002), or pollination (Tscheulin et al., 2011). Effort should be made in the assessment of patches of natural vegetation surrounding the farm and in the landscape composition if the objective is to enhance the conservation biological control in olive groves.

References

Agustí, N., Unruh, T. R. & Welter, S. C. (2003). Detecting *Cacopsylla pyricola* (Hemiptera: Psyllidae) in predator guts using COI mitochondrial markers. *Bulletin of Entomological Research*, *93*, 179-185.

Aldebis, H. K., Avila, A., Mata, M. & Vargas-Osuna, E. (2004). Evaluation of the damage caused by the olive moth, *Prays oleae* Bern. in different varieties and growing conditions. *Pest Plant Protection Bulletin*, *30*, 649-656.

Altieri, M. A. (1991). How best can we use biodiversity in agroecosystems. *Outlook on Agriculture*, *20*, 15-23.

Altieri, M. A., Ponti, L. & Nicholls, C. I. (2005). Manipulating vineyard biodiversity for improved insect pest management: case studies from northern California. *International Journal of Biodiversity Science and Management*, *1*, 1-13.

Alvarado, M., Civantos, M. & Durán, J. M. (2008). Plagas. In: D. Barranco & R. Fernández-Escobar, L. Rallo (Eds.) El Cultivo del Olivo. (pp. 509-593). Madrid: Mundi Prensa.

Arambourg, Y. (1986). Traite d'Entomologie Oleicole. Madrid: Consejo Oleícola Internacional.

Barbosa, P. (1998). Conservation Biological Control. London: Academic press.

Begum, M., Gurr, G. M., Wratten, S. D., Hedberg, P. R. & Nicol, H. I. (2006). Using selective food plants to maximize biological control of pests vineyards. *Journal of Applied Ecology*, *43*, 547-554.

Bianchi, F., Booij, C. J. H. & Tscharntke, T. (2006). Sustainable pest regulation in agricultural landscapes: a review on landscape composition, biodiversity and the natural pest control. *Proceedings of the Royal Society B: Biological Sciences*, *273*, 1715-1727.

Blitzer, E. J., Dormann, C. F., Holzschuh, A., Klein, A. M., Rand, T. A. & Tscharntke, T. (2012). Spillover of functionally important organism between managed and natural habitats. *Agriculture, Ecosystem and Environment*, *146*, 34-43.

Boccaccio, L., Petacchi, R. (2009). Landscape effects on the complex of *Bactrocera oleae* parasitoids and implications for conservation biological control. *BioControl*, *54*, 607-616.

Bone, N. J., Thomson, L. J., Ridland, P. M., Cole, P. & Hoffmann, A. A. (2009). Cover crops in Victorian apple orchards: Effects on production, natural enemies and pests across a season. *Crop Protection*, *28*, 675-683.

Campos, M. (2011). Las plagas y su manejo en el olivar ecológico. In: M. A. Alonso et al. (Eds.) *El olivar ecológico* (pp. 155-201). Sevilla: Mundi-Prensa.

Cárdenas, M. (2008). Análisis de la diversidad ecológica de las arañas en el olivar. PhD Thesis. Granada: University of Granada.

Civantos, M. (1999). *Control de Plagas y Enfermedades del Olivar*. Madrid: Consejo Oleícola Internacional.

Chaplin-Kramer, R., Rourke, M. E., Blitzer, E. J. & Kremen, C. (2011). A meta-analysis of crop pest and the natural enemy response to landscape complexity. *Ecology Letters*, *14*, 922-932.

Colloff, M. J., Lindsay, E. A. & Cook, D. C. (2013). Natural pest control on citrus as an ecosystem service: integrating ecology, economics and management at the farm scale. *Biological Control*, *67*, 170-177.

Cotes, B., Campos, M., Pascual, F. & Ruano, F. (2010). The ladybeetle community (Coleoptera: Coccinellidae) in southern olive agroecosystems of Spain. *Environmental Entomology*, *39*, 79-87.

Croft, B. A. & Brown, W. A. (1975). Responses of arthropods natural enemies to insecticide. *Annual Review of Entomology*, *20*, 285-335.

Cullen, R., Warner, K. D., Jonsson, M. & Wratten, S. D. (2008). Economics and adoption of conservation biological control. *Biological Control*, *45*, 272-280.

Danne, A., Thomson, L. J., Sharley, D. J., Penfold, C. M. & Hoffmann, A. A. (2010). Effects of native grass cover crops on beneficial and pest invertebrates in Australian vineyards. *Environmental Entomology*, *39*, 970-978.

Denoth, M., Frid, L. & Myers, J. H. (2002). Multiple agents in biological control: improving the odds? *Biological Control*, *24*, 20-30.

Drukker, B., Scutareanu, P. & Sabelis, M. W. (1995). Do anthocorid predators respond to synomones from *Psylla*-infested trees under field conditions? *Entomologia Experimentalis et Applicata*, *77*, 193-203.

Duarte, J., Campos, M., Guzmán, J. R., Beaufoy, G., Farfan, M. A., Cotes, B., Benítez, E., Vargas, J. M. & Muñoz-Cobo, J. (2009). Olivar y biodiversidad. In: Guzmán-Alvares, J. R., Gómez, J. A. & Rallo L. *Sostenibilidad de la Producción de Olivar en Andalucía* (pp. 119-149). Sevilla: Consejería de Agricultura y Pesca.

FAOSTAT. (2014). Statistics of the Food and Agriculture Organization. Available at: http://faostat3.fao.org/browse/Q/QC/E.

Hardman, J. M., Franklin, J. L., Bostanian, N. J. & Thistlewood, H. M. A. (2011). Effect of the width of the strip on herbicide in apple orchards mite dynamics. *Experimental and Applied Acarology*, *53*, 215-234.

Hartwig, N. L. & Ammon, H. U. (2002). 50th Anniversary - Invited article: Cover crops and living mulches. *Weed Science*, *50*, 688-699.

Horton, D. R. & Lewis, T. M. (2000). Seasonal distribution of *Anthocoris* spp. and *Deraeocoris brevis* (Heteroptera: Anthocoridae, Miridae) in orchard and non-orchard habitats of central Washington. *Annals of the Entomological Society of America*, *93*, 476-485.

Landis, D. A., Wratten, S. D. & Gurr, G. M. (2000). Habitat management to conserve natural enemies of arthropod pests in agriculture. *Annual Review of Entomology*, *45*, 175-201.

Logan, J. D., Wolesensky, W. & Joern, A. (2006). Temperature-dependent phenology and predation in arthropods systems. *Ecological Modelling, 196*, 471-482.

Loureau, M. (1998). Biodiversity and ecosystem functioning: a mechanistic model. *Proceedings of the National Academy of Science USA*, *95*, 5632-5636.

Karp, D. S., Mendenhall, C. D., Sandí, S. F., Ehrlich, P. R., Hadly, E. A. & Daily, G. C. (2013). Forest bolsters birds, pest control, and coffee yield. *Ecology Letters*, *16*, 1339-1347.

Martin, E. A., Reinekin, B., Seo, B. & Steffan-Dewenter, I. (2013). Natural enemy interactions constrain pest control in complex agricultural landscapes. *Proceedings of the National Academy of Science USA*, *110*, 5534-5539.

Martinat, P. J. (1987). The role of climatic variation and weather in forest insect outbreaks. In: P. Barbosa & J. C. Schultz (Eds). *Insect Outbreaks*. (pp. 241- 268). New York: Academic Press Inc.

Meehan, T. D., Werling, B. P., Landis, D. A. & Gratton, C. (2011). Agricultural landscape simplification and insecticide use in the Midwestern United States. *Proceedings of the National Academy of Science USA*, *108*, 11500-11505.

Moonen, A. C. & Bàrberi, P. (2008). Functional biodiversity: An agroecosystem approach. *Agriculture, Ecosystem and Environment*, *127*, 7-21.

Morris, T. I., Campos, M., Kidd, N. A. C. & Symondson, W. O. C. (1999). What is consuming *Prays oleae* (Bernard) (Lep.: Yponomeutidae) and when: a serological solution? *Crop Protection 18*, 17–22.

Ortega, M. & Pascual, S. (2014). Spatio-temporal analysis of the relationship between landscape structure and the olive fruit fly *Bactrocera oleae* (Diptera: Tephritidae). *Agricultural and Forest Entomology*, *16*, 14-23.

Paredes, D., Campos, M. & Cayuela, L. (2013a). El control biológico de plagas de artrópodos por conservación: técnicas y estado del arte. *Ecosistemas*, *22*, 58-63.

Paredes, D., Cayuela, L. & Campos, M. (2013b). Synergistic effects of ground cover and surrounding natural vegetation on natural enemies of olive pests. *Agriculture, Ecosystems and Environment*, *173*, 72-80.

Paredes, D., Cayuela, L., Gurr, G. M. & Campos, M. (2013c). Effect of non-crop vegetation types on conservation biological control of pests in olive groves. *PeerJ*, *1*, e116.

Paredes, D., Cayuela, L., Gurr, G. M. & Campos, M. (2015a). Is ground cover vegetation an effective biological control strategy against olive pests? *PLoS ONE*, *10*, e0117265.

Paredes, D., Cayuela, L., Gurr, G. M. & Campos, M. (2015b). Single best species or natural enemy assemblages? A correlational approach to investigating ecosystem function. *BioControl*, *60*, 37-45.

Perdikis, D., Fantinou, A. & Lykouressis, D. (2011). Enhancing pest control in annual crops by conservation of predatory Heteroptera. *Biological Control*, *59*, 13-21.

Perovic, D. J., Gurr, G. M., Raman, A. & Nicol, H. I. (2010). Effect of landscape composition and arrangement on biological control agents in a simplified agricultural system: A cost-distance approach. *Biological Control*, *52*, 263-270.

Polis, G. A. & Holt, R. D. (1992). Intraguild predation: the dynamics of complex trophic interactions. *Trends in Ecology and Evolution*, *7*, 151-154.

Redolfi, I., Tinaut, A., Pascual, F. & Campos, M. (1999). Qualitative aspects of myrmecocenosis (Hymenoptera: Formicidae) in olive orchards with different agricultural management in Spain. *Journal of Applied Entomology*, *123*, 621-627.

Rei, F., Campos, M. & Torres, L. (2010). Beneficial arthropods associated with the olive tree canopy in Alentejo (South region of Portugal). *IOBC/wprs Bulletin*, *53*, 117-127.

Rieux, R., Simon, S. & Defrance, H. (1999). Role of hedgerows and ground cover management on arthropod populations in pear orchards. *Agriculture, Ecosystems and Environment*, *73*, 119-127.

Rodríguez, E., González, B. & Campos, M. (2009). Effects of cereal cover crops on the main insect pests in Spanish olive orchards. *Journal of Pest Science*, *82*, 179-185.

Ruiz-Torres, M. & Montiel-Bueno, A. (2000). Introducción al conocimiento de la entomofauna del olivar en la provincia de Jaén. Aspectos cualitativos (I). *Boletín de sanidad vegetal. Plagas*, *26*, 129-147.

Rusch, A., Bommarco, R., Jonsson, M., Smith, H. G. & Ekbom, B. (2013). Natural flow and stability of pest control services depend on complexity and crop rotation at the landscape scale. *Journal of Applied Ecology*, *50*, 345-354.

Rusch, A., Chaplin-Kramen, R., Gardiner, M. M., Hawro, V., Holland, J., Landis, D., Thies, C., Tscharntke, T., Weisser, W. W., Winqvist, C., Woltz, M. & Bommarco, R. (2016). Agricultural landscape simplification reduces natural pest control: A quantitative synthesis. *Agriculture, Ecosystem and Environment*, *221*, 198-204.

Santos, S. A. P., Cabanas, J. E. & Pereira, J. A. (2007). Abundance and diversity of soil arthropods in olive grove ecosystem (Portugal): Effect of pitfall trap type. *European Journal of Soil Biology*, *43*, 77-83.

Shaltiel, L. & Coll, M. (2004). Reduction of pear psylla damage by the predatory bug *Anthocoris nemoralis* (Heteroptera: Anthocoridae): The importance of orchard colonisation time and neighbouring vegetation. *Biocontrol Science and Technology*, *14*, 811-821.

Silva, E. B., Franco, J. C., Vasconcelos, T. & Branco, M. (2010). Effect of ground cover vegetation on the abundance and diversity of beneficial arthropods in citrus orchards. *Bulletin of Entomological Research*, *100*, 489-499.

Simon, S., Bouvier, J. C., Debras, J. F. & Sauphanor, B. (2010). Biodiversity and pest management in orchard systems: A review. *Agronomy for a Sustainable Development*, *30*, 139-152.

Smith, M. W., Arnold, D. C., Eikenbary, R. D., Rice, N. R., Shiferaw, A., Cheary, B. S. & Carroll, B. L. (1996). Influence of ground cover on beneficial arthropods in pecan. *Biological Control*, *6*, 164-176.

Song, B. Z., Wu, H. Y., Kong, Y., Zhang, J., Du, Y. L., Hu, J. H. & Yao, Y. C. (2010). Effects of intercropping with aromatic plants on the diversity and structure of an arthropod community in a pear orchard. *BioCcontrol*, *55*, 741-751.

Straub, C. S., Finke, D. L. & Snyder, W. E. (2008). Are the nature conservation of biodiversity and biological control enemy compatible goals? *Biological Control*, *45*. 225- 237.

Thies, C. & Tscharntke, T. (1999). Landscape structure and biological control in agroecosystems. *Science*, *285*, 893-895.

Tscheulin, T., Neokosmidis, L., Petanidou, T. & Settele, J. (2011). Influence of landscape context on the abundance and diversity of bees in Mediterranean olive groves. *Bulletin of Entomological Research*, *101*, 557-564.

Tsitsilas, A., Hoffman, A. A., Weeks, A. R. & Umina, P. A. (2011). Impact of groundcover manipulations within windbreaks on mite pests and their natural enemies. *Australian Journal of Entomology*, *50*, 37-47.

Tzanakakis, M. (2006). Insects and Mites Feeding on Olive. Distribution, Importance, Habits, Seasonal Development, and Dormancy. Leiden-Boston: Brill.

Walker, B. H. (1992). Biodiversity and ecological redundancy. *Conservation Biology*, *6*, 18–23.

Woltz, J. M., Isaacs, R. & Landis, D. A. (2012). Landscape structure and habitat management differentially influence insect natural enemies in an agricultural landscape. *Agriculture, Ecosystems and Environment*, *152*, 40-49.

Biographical Sketches

Daniel Paredes Llanes, PhD

Departamento de Protección Ambiental.

Estación Experimental del Zaidín.

Consejo Superior de Investigaciones científicas.

Research and Professional Experience: His research focuses on the study of ecosystem services in agricultural landscapes.

Education: Ph.D. Universidad Granada, Spain

Education at other institutions: MSc, Universidad de Granada; BSc Universidad de Extremadura, Spain

Honors: Cum Laude Doctor Europeus

Publications Last Three Years:

Paredes, D.; Moreno-Chocano, J.; Cano, A.; Ruano, F.; Campos, M.. (2015). Estudio de fecundidad y ciclo de vida de *Anthocoris nemoralis* (Heteroptera: Anthocoridae) en el olivar. *Revista de Fruticultura 44*, 18 – 25.

Paredes, D.; Cayuela, L.; Gurr, G.M.; Campos, M. (2015). Is ground cover vegetation an effective biological control enhancement strategy against Olive Pests? *PLoS ONE, 10.*

Paredes, D.; Cayuela, L.; Gurr, G.M.; Campos, M. (2014). Single best species or natural enemy assemblages? a correlational approach to investigating ecosystem function. *BioControl 60*, 37- 45.

Paredes, D.; Campos, M. (2014). La vegetación nativa y el control de plagas en el olivar. *Revista de Fruticultura, 34*, 2-8.

Paredes, D.; Cayuela, L.; Gurr, G.M; Campos, M. (2013). Effect of non-crop vegetation types on conservation biological control of pests in olive groves. *PeerJ.*

Paredes, D.; Cayuela, L.; Campos, M. (2013). Synergistic effects of ground cover and adjacent vegetation on natural enemies of olive insect pests. *Agriculture, Ecosystems and Environment*, 173, 72- 80.

Paredes, D.; Campos, M. (2013). Importancia de la biodiversidad en la gestión integrada de plagas en olivar. *Vida Rural, 363*, 30-34.

Paredes, D.; Campos, M.; Cayuela, L. (2013). El control biológico de plagas de artrópodos por conservación: técnicas y estado del arte. *Ecosistemas 22*, 58 – 63.

Luis Cayuela Delgado, PhD

Área biodiversidad y conservación. Universidad Rey Juan Carlos I

Research and Professional Experience: He is interested in forest ecology, conservation biology, agroecology and environment statistics developed under R software.

Professional Appointments: Associated Professor

Education: Ph.D. Universidad Alcalá, Spain

Education at other institutions: MSc in Conservation, University College of London (United Kingdom)

Honors: European PhD and Cum Laude

Publications Last Three Years:

Albuquerque, F.S.; Assunção-Albuquerque, M.J.T.; Cayuela, L.; Zamora, R.; Benito, B.M. (2013). European bird distribution is "well" represented by special protected areas: mission accomplished? *Biological Conservation* 159: 45-50.

Albuquerque, F.S.; Benito, B.; Beier, P.; Assunção-Albuquerque, M.; Cayuela, L. (En prensa). Supporting underrepresented forests in Mesoamerica. *Natureza & Conservação*.

Altamirano, A.; Aplin, P.; Miranda, A.; Cayuela, L.; Algar, A.; Field, R. (2013). High rates of forest loss and turnover obscured by classical landscape measures. *Applied Geography* 40: 199-211.

Benito, B.M.; Cayuela, L.; Albuquerque, F.S. (2013). The impact of modeling choices in the predictive performance of diversity maps derived from species distribution models: guidelines to build better models. *Methods in Ecology and Evolution* 4: 327-335.

Cayuela, L.; Gotelli, N. J.; Colwell, R. K. (2015). Ecological and biogeographical null hypotheses for comparing rarefaction curves. *Ecological Monographs* 85: 437-455.

Cayuela, L.; Hernández, R.; Hódar, J. A.; Sánchez, G.; Zamora, R. (2014). Tree damage and population density relationships for the pine processionary moth: prospects for ecological research and pest management. *Forest Ecology and Management* 328: 319-325.

De Miguel, R.; Gálvez-Bravo, L.; Oliva-Paterna, F.; Fernández-Delgado, C.; Cayuela, L. (En prensa). Recolonization process and fish assemblage dynamics in the Guadiamar River (SW Spain) after the Aznalcóllar mine toxic spill. *River Research and Applications*.

Espinosa, C.; Cayuela, L. (2015). Biodiversidad escondida. *Investigación y Ciencia* 466 (julio): 52-53.

Garibaldi, C.; Nieto-Ariza, B.; Macía, M.; Cayuela, L. (2014). Soil and geographical distance as determinants of floristic composition in the Azuero peninsula. *Biotropica* 46: 687-695.

Ledo, A.; Cayuela, L.; Manso, R.; Condés, S. (2015). Recruitment patterns and potential mechanisms of community assembly in an Andean Cloud Forest. *Journal of Vegetation Science* 26: 876-888.

Miranda, A.; Altamirano, A.; Cayuela, L.; Pincheira, F.; Lara, A. (2015) Different times, same history: native forest loss and landscape homogenization in three physiographical areas of south-central Chile. *Applied Geography* 60: 20-28.

Newton, A. C.; Oldfield, S.; Rivers, M.; Mark, J.; Schatz, G.; Tejedor-Garavito, N.; Cantarello, E.; Golicher, D.; Cayuela, L.; Miles, L. (2015). Towards a global tree assessment. *Oryx* 49: 410-415.

Paredes, D.; Campos, M.; Cayuela, L. (2013). El control biológico de plagas de artrópodos por conservación: técnicas y estado del arte. *Ecosistemas* 22: 56-61.

Paredes, D.; Cayuela, L.; Campos, M. (2013). Synergistic effects of ground cover and adjacent vegetation on natural enemies of olive insect pests. *Agriculture, Ecosystems and Environment* 173: 72-80.

Paredes, D.; Cayuela, L.; Gurr, G. M.; Campos, M. (2015). Is ground cover vegetation an effective biological control enhancement strategy against olive pests? *PLoS ONE* 10(2): e0117265.

Paredes, D.; Cayuela, L.; Gurr, G. M.; Campos, M. (2015). Single best species or natural enemy assemblages? A correlational approach to investigating ecosystem function. *BioControl* 60: 37-45.

Paredes, D.; Cayuela, L.; Gurr, G.M.; Campos, M. (2013). Effect of non-crop vegetation types on conservation biological controlof pests in olive groves. *PeerJ*: e116.

Zamorano-Elgueta, C.; Cayuela, L.; Rey-Benayas, J.M.; Donoso, P.J.; Geneletti, D.; Hobbs, R.J. (2014). The differential influences of human-induced disturbances on tree regeneration community: a landscape approach. *Ecosphere* 5: 90.

Zamorano-Elgueta, C.; Rey-Benayas, J. M.; Cayuela, L.; Hantson, S.; Armenteras, D. (2015). Native forest replacement by exotic plantations in southern Chile (1985-2011) and partial compensation by natural regeneration. *Forest Ecology and Management* 345: 10-20.

Mercedes Campos Aranda, PhD

Departamento de Protección Ambiental. Estación Experimental del Zaidín. Consejo Superior de Investigaciones Científicas

Research and Professional Experience: She is the head of the crop protection group at Spanish Council of Research and has and strong background in biological control of pest in olive groves.

Professional Appointments: Professor of Research

Education: Ph. D. Universidad Granada

Honors: Cum Laude

Publications Last Three Years:

Pascual, J.; Blanco, S.; García-López, M.; García-Salamanca, A.; Bursakov, S.A.; Genilloud, O.; Ramos, J.L.; van Dillewijn, P. (2016). Assessing bacterial diversity in the rhizosphere of *Thymus zygis* growing in the

Sierra Nevada National Park (Spain) through culture-dependent and independent approaches. PLoS ONE, 11(1): e0146558

Santos, S.A.P.; Mota, L.; Malheiro, R.; Silva, F.; Campos, M.; Guedes de Pinho, P.; Pereira, J.A. (2016). Changes in volatile compounds of *Dittrichia viscosa* caused by the attack of the gall-forming diptean *Myopites stylatus*. *Industrial Crops and Products*, 87: 71-77

Gonzalez, D.; Nave, A.; Gonçalves, F.; Nunes, F.M.; Campos, M.; Torres, L. (2016). Effects of ten naturally occurring sugars on the reproductive success of the green lacewing, *Chrysoperla carnea. BioControl*, 61: 57-67

Paredes, D.; Moreno-Chocano, J.; Cano, A.; Ruano, F.; Campos, M. (2015). Estudio de fecundidad y ciclo de vida de Anthocoris nemoralis (Heteroptera: Anthocoridae) en el olivar. *Revista de Fruticultura*, 18-25

Cárdenas, M.; Pascual, F.; Campos, M.; Pekár, S. (2015). The spider assemblage of olive groves under three management systems. *Environmental Entomology*, 44: 509-518

Paredes, D.; Cayuela, L.; Gurr, G.M.; Campos, M. (2015). Is ground cover vegetation an effective biological control enhancement strategy against Olive Pests?. *PLoS ONE,* 10(2): e0117265.

Paredes, D.; Cayuela, L.; Gurr, G.M.; Campos, M. (2014). Single best species or natural enemy assemblages? a correlational approach to investigating ecosystem function. *BioControl* 60:37-45.

Paredes, D.; Campos, M.. (2014). La vegetación nativa y el control de plagas en el olivar. Revista de *Fruticultura*, 34:2-8.

Molina, L.; Udaondo, Z.; Duque, E.; Fernández, M.; Molina-Santiago, C.; Roca, A.; Porcel, M.; De La Torre, J.; Segura, A.; Plesiat, P.; Jeannot, K.; Ramos, J.-L. (2014). Antibiotic resistance determinants in a *Pseudomonas putida* strain isolated from a hospital. *PLoS ONE*: 9(1): e81604

Jerez-Valle, C.; García, P.A.; Campos, M.; Pascual, F.. (2014). A simple bioindication method to discriminate olive orchard management types using the soil arthropod fauna. *Applied Soil Ecology* 76: 42-51.

D. Paredes; L. Cayuela; G.M. Gurr; M. Campos. (2013). Effect of non-crop vegetation types on conservation biological control of pests in olive groves. *Peer J. 1:e116.*

Paredes, D.; Cayuela, L.; Campos, M. (2013). Synergistic effects of ground cover and adjacent vegetation on natural enemies of olive insect pests. *Agriculture, Ecosystems and Environment* 173:72-80.

Paredes, D.; Campos, M. (2013). Importancia de la biodiversidad en la gestión integrada de plagas en olivar. *Vida Rural* 36:30-34.

Paredes, D.; Campos, M.; Cayuela, L. (2013). El control biológico de plagas de artrópodos por conservación: técnicas y estado del arte. *Ecosistemas* 22: 58-63.

Porcel. M.; Ruano, F.; Cotes, B.; Peña, A.; Campos, M. (2013). Agricultural management systems affect the green lacewing community (Neuroptera: Chrysopidae) in olive orchards in southern Spain. *Environmental Entomology* 42:97-106.

In: Natural Enemies
Editor: Sónia A.P. Santos

ISBN: 978-1-63485-921-9

Chapter 2

BIOLOGICAL PEST CONTROL OF THE TETRANYCHID MITE *OLIGONYCHUS PERSEAE* IN AVOCADO AND CHALLENGES TO FACE IN A CLIMATE CHANGE SCENARIO

Celeste Guzmán*[*] *and Marta Montserrat
Departamento de Fruticultura Subtropical,
Instituto de Hortofruticultura Subtropical y Mediterránea "La Mayora",
Universidad de Málaga-Consejo Superior de Investigaciones Científicas
(IHSM-UMA-CSIC), Algarrobo-Costa, Málaga, Spain

ABSTRACT

The avocado (*Persea americana* Mill.) is a fruit tree native to Mesoamerica that experienced an important expansion to adjacent regions in Central and South America, probably because of the commercial exchange between native civilizations. Following the arrival of the European colonizers in 1492, the avocado began to expand to the five continents. In 2013, the total world avocado harvested area was 516.485 ha with a production of 4.717.103 tons (FAOSTAT), indicating the great social and economic importance of this crop. Avocado production is undermined by attacks of several insects and acari. In this chapter, we

[*] Corresponding author: Functional and Evolutionary Ecology, Estación Experimental de Zonas Áridas, CSIC. Carretera de Sacramento s/n. La Cañada De San Urbano, Almería, Spain. Email: cguzman@eeza.csic.es.

focused on one of the most important pests affecting avocado crops in producer countries such as Mexico, Costa Rica, the USA, Israel and Spain, the persea mite *Oligonychus perseae* (Tuttle, Baker and Abbatiello). This herbivore mite, specific to avocados, probably co-evolved with its host, as both come from the same area of origin. *O. perseae* colonizes avocado leaves, preferentially the underside, building semi-circular dense silken nests along the midrib and main veins. Nests protect individuals against some predators and adverse abiotic conditions, providing refuge for their development, reproduction and feeding. This herbivore sucks out the cellular content of the leaf tissues, causing necrotic spots that damage the leaves, affecting the photosynthetic capacity and, in severe infestations, causing tree defoliation. Several species of predatory mites of the Phytoseiidae family have been identified as natural enemies of this pest: *Euseius* spp (*E. stipulatus* (Athias-Henriot) and *E. scutalis* (Athias-Henriot)), *Neoseiulus californicus* (McGregor), and *Galendromus* spp (*G. annectens* (De Leon) and *G. helveolus* (Chant)), among others. Individuals of these species were studied and later used as biocontrol agents in augmentative and conservative biological control strategies in avocado orchards in different countries. However, such biocontrol strategies are facing disruptive impacts caused by the ongoing climate change. Increasing temperatures induce multiple shifts in the structure and dynamics of natural and agricultural communities, because they are affecting life-history parameters, especially in ectotherm organisms such as arthropods, and the ecological interactions occurring between them. Indeed, several experimental works in avocado and in other crops have found that global warming affects agricultural trophic networks in a way that often favours herbivory. It should be noted that determining the effects that the current climate change is causing on agricultural systems is a complex objective. On the one hand, changes in abiotic conditions (e.g., temperature and precipitation) are heterogeneous and vary across regions. On the other hand, such effects are highly variable due to the peculiarities of each system, which make them unique in their performance and in their response against environmental disruptions. Therefore, an extensive knowledge of every system is essential and a big effort in fundamental and applied research necessary, in order to be able to predict how communities will respond to actual and future climate conditions, and to build successful biological pest control (BPC) programs in different agroecosystems.

Keywords: avocado, *Oligonychus perseae*, phytoseiids, biological control, climate change

1. Climate Change and Its Implication on Natural and Agricultural Systems

Scientists and experts of several disciplines from around the world have participated in The Intergovernmental Panel on Climate Change (IPCC), since its creation in 1988 by the World Meteorological Organization (WMO) and the United Nations Environment Program (UNEP). The objective of this initiative is to join the available independent scientific information and analyses regarding all aspects related with the ongoing climate change, providing to the governments and policymakers the tools to face and mitigate its potential effects and impacts at environmental, social and economic levels. Based on this knowledge, IPCC stated in their latest reports (2007 and 2014) that climate change is undeniable. There are empirical evidences of the increase in global atmosphere and ocean average temperatures, resulting in a generalized melting of ice, a rise of sea levels and an increase in the amount of greenhouse gases (IPCC, 2014). The global warming observed since the mid-20th century is extremely likely (95-100% probability) caused by the human influence, and is currently affecting natural systems and human activity in all continents (IPCC, 2014). On a minor scale, shifts in temperatures and precipitations will vary among the different regions, so we have to keep in mind that the species will respond in a different way based on the local variations of the environmental conditions they are experiencing.

In the literature we can find a multitude of evidences and experimental studies that confirm that climate change is generating shifts in the physiology and performance of individuals and altering all aspects of communities (Cock et al., 2013; IPCC, 2014; Thomson et al., 2010), including phenology and synchrony between species, population abundances, species range distribution, diversity, and composition, migration patterns, causing the extinction of species and affecting biotic interactions as competition, predation, parasitism, mutualism, etc. which are commonly climate-dependent (Beveridge et al., 2010; Dunson & Travis, 1991; Gilman et al., 2010). Tylianakis et al. (2008) presented a review of 688 publications that demonstrated evidences of the effect caused by the main drivers of the global environmental change in the strength and/or the direction of biotic interactions, concluding that these effects are very important but both highly variable depending on the specific system. For instance, the mutualistic relationship between plants and their pollinators could be negatively affected if climate change produces phenological mismatches between the flowering period and the pollinator activity (Memmott et al., 2007). Another example is the increase in temperatures which may favor the establishment of invasive species that could cause the exclusion and replacement of native ones through competition for food, space, etc. predation or transmission of diseases (Snyder & Evans, 2006), altering the structure and composition of the community.

In addition to those examples mentioned above, the agricultural systems suffer other potential effects of climate change, because global warming provides climate corridors that may facilitate the arrival of some invasive species that can act as pests in their new environment, promoting an increase in both the diversity and abundance of phytophagous pests (Cock et al., 2013). Another consequence of changes in temperatures is that sensitivity to warming increases with the trophic level (Preisser & Strong, 2004; Voigt et al., 2003), being more negative to predators than to prey (pests), probably resulting in an increase of herbivory (Montserrat et al., 2013a). In agricultural systems handled by biological pest control (BPC), climate change can induce several effects on the performance of natural enemies (predators and parasitoids). These are complex effects as they can be positive or negative for the BPC, depending on the particular system, and may even be difficult to determine. For instance, Coll & Hughes (2008) documented a case in which one driver of climate change could favour the BPC: in an environment with elevated CO_2, pea plants (*Pisum sativum* L.) presented reduced nitrogen content, thus the *Helicoverpa armigera* (Hübner) larvae that fed on them were smaller than the ones that fed on pea plants with atmospheric CO_2. Consequently, the predator of these larvae, the pentatomid bug *Oechalia schellenbergii* (Guérin-Méneville), was more effective due to the higher vulnerability of the prey. However, in general, it is expected that climate change (and specifically global warming) will negatively affect BPC as the majority of natural enemies involved are arthropods, i.e., ectotherm organisms that cannot regulate their own temperature, therefore being highly dependent and susceptible, to environmental conditions. The development, movement capacity, speed and predation rate of these organisms depend on abiotic conditions. Hence, the strength of the interactions between individuals (predators-prey, predators-predators) will vary if abiotic conditions change, in turn, affecting the effectiveness of BPC in crops under climate change.

Populations of herbivores are also affected by abiotic conditions, but to a lesser extent than their predators. Roy et al. (2003) determined the effect of 12 temperature regimes (between 10–38°C) on the intrinsic rates of natural increase (r_m) of a pest of raspberries, the spider mite *Tetranychus mcdanieli* McGregor, and of its predator, the coccinellid *Stethorus punctillum* Weise.

Results showed that above 24ºC the survival was still high for the pest, but had considerably decreased for the predator. Moreover, in the suitable range for population growth of both species, the r_m of the pest *T. mcdanieli* was 1.9 to 8 times greater than the r_m of its predator *S. punctillum*. They concluded that, based on the theory, as this predator does not present a r_m close to, or higher than, the r_m of its prey, it may not be a suitable biocontrol agent to control *T. mcdanieli* populations (Roy et al., 2003).

Other abiotic factors that can influence the BPC are the availability of water and the relative humidity (RH). Cagan et al. (1998) detected that the extremely dry weather during 1993 in Slovakia caused the parasitoid *Trichogramma* to fail in parasiting *Ostrinia nubilalis* (Hübner) eggs. IPCC also predicts an increase in the intensity, frequency and duration of extreme weather events such as heat waves, severe droughts, hurricanes and floods. There is a risk of pest population outbreaks after these extreme events (Cock et al., 2013) due to, among other causes, the effects on natural enemies. For instance, the least reduction of the spider mite *Tetranychus urticae* Koch by the phytoseiid predator *Phytoseiulus persimilis* Athias-Henriot was found in bean plants exposed to severe drought stress, in comparison to moderately stressed plants (English-Loeb, 1990).

Lastly, farmers may carry out alterations of their practices, adopting different management strategies to confront the climate change effects (shifts in irrigation, adoption of no-tillage plantings, use of mulches, changes in crop varieties, implementation of revegetation programs, etc.) consequently causing variations in abundance, distribution, activity and interactions on natural enemies (Thomson et al., 2010).

2. The Mite Community on Avocado

The avocado (*Persea americana* Mill., Lauraceae Family) is a fruit tree native to Mesoamerica, specifically from the highlands of Mexico and Guatemala. It is believed that Mexican avocados have been used by Indians since 7800 B.C. approximately, as some archeological studies of fossils found in caves in Oaxaca demonstrated (Williams, 1977). The avocado domestication occurred in its area of origin and expanded to Centroamerica (Colombia, Venezuela, Ecuador and Peru), probably because of the commercial exchange between native civilizations. Following the arrival of the European colonizers in 1492, the avocado began to expand to the rest of the world, firstly reaching Spain in 1600 (Smith et al., 1992) and is currently present in all five continents.

The avocado, the fourth most important tropical fruit in the world, has a multitude of uses, preferentially in gastronomy and nutrition, and incorporated in cosmetics and natural medicines, which has increased its demand. In 2013, the total world avocado harvested area was 516.485 ha, with a production of 4.717.103 tons. Some of the producer countries are Mexico with 1.467.837 tons, Peru with 288.387 tons, the USA with 175.226 tons, Israel with 91.904 tons and Spain with 69.400 tons, among others (FAOSTAT).

In regards to pests, avocado production suffers from the attack of several insects and acari. One of the most important herbivores affecting avocado crops, preferentially Hass, Gwen and Reed varieties, is the persea mite *Oligonychus perseae* (Tuttle, Baker and Abbatiello) (Bender, 1993). It was detected in the USA (Aponte & McMurtry, 1997; Bender, 1993; Hoddle & Morse, 2013), in Costa Rica (Salas, 1978), in Israel (Maoz et al., 2011), in Portugal (Borges et al., 2008), in Spain (González-Fernández et al., 2009; Vela et al., 2007) and recently in Italy (Zappalà et al., 2015). This tetranychid mite specific to avocados, may also attack other plants such as *Malva* spp., grapes, roses, deciduous fruits such apricots, nectarines and peaches and some trees, namely *Salix* sp., carob and annatto (Aponte & McMurtry, 1997; Bender, 1993). In avocado, *O. perseae* colonizes the leaves, preferentially the abaxial surface, building semi-circular dense silken nests along the midrib (Figure 1), main veins and leaf depressions, and, when population densities are very high, the individuals can disperse to the upper side of the leaves. Nests protect individuals against some predators and adverse abiotic conditions, providing refuge for their development, reproduction and feeding. The mites use the veins and midrib as anchor points, from which they use silken strands to construct the nest in a zig-zag pattern, letting one or more semicircular gaps, mainly at margins, to be used as their entrance and exit (Aponte & McMurtry, 1997). This herbivore sucks out the cellular content of the leaf tissues, causing necrotic spots that damage the leaves, affecting the photosynthetic capacity and, in severe infestations (from 500 mites per leaf), causing tree defoliation and consequently increasing the risk of sunburn to fruits and tree trunks (Bender, 1993).

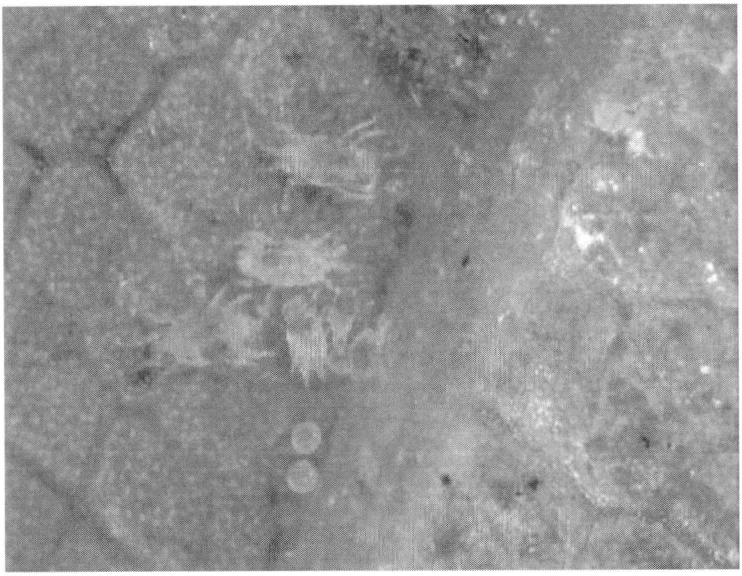

Figure 1. *Oligonychus perseae* nest alongside the midrib with adults, mobile juveniles, and eggs.

There are several species of natural enemies that maintain *O. perseae* populations at non-risk or sub-economic injury levels, either by natural or artificial (augmentative or conservative) control. In particular, predatory mites of the Phytoseiidae family have been identified as natural enemies of this pest: *Euseius stipulatus* (Athias-Henriot), *E. scutalis* (Athias-Henriot), *Neoseiulus californicus* (McGregor), *Galendromus annectens* (De Leon) and *G. helveolus* (Chant), among others. *E. stipulatus* and *E. scutalis* are omnivores, with a diet based on the pollen of several plants and on tetranychid and eriophyid preys (classified as Type IV specialized pollen feeders/generalist predators in McMurtry and Croft (1997)). *N. californicus, G. annectens* and *G. helveolus* are specific predators of tetranychid mites (Type II in McMurtry and Croft (1997)). *N. californicus* and *G. helveolus* can penetrate inside the nest of *O. perseae*, the first one using its forelegs to rip and open the nest (Montserrat et al., 2008), and due to its small size, the second one is able to penetrate through the entrance of the nest. Moreover, both species are commercially available, making them suitable agents for *O. perseae* augmentative biocontrol strategies.

3. *Oligonychus Perseae* Biocontrol Strategies in Different Countries

O. perseae was first detected in the United States, in California, in 1990 (Bender, 1993), where it had probably arrived through infested avocado plant material from Mexico, because at that time the control measures and detection of potential pests had not been as extensive as when the United States strengthened its preventive measures, including the quarantine of incoming avocados (Lara & Hoddle, 2015a). In California, the highest risk of damage associated with this herbivore occurs in the summer growing season, when populations reach their maximum densities (Lara & Hoddle, 2015a), so experts have developed accurate sampling plans with guidelines for growers to monitor *O. perseae* populations, in order for them to take the best decisions related with biocontrol strategies (Lara & Hoddle, 2013, 2015b; Lara et al., 2016). In the avocado orchards of California, an abundant and widely distributed natural enemy of this pest is the native species *Euseius hibisci* (Chant). However, several studies propound that this predator mite is not an effective natural enemy to be used as biocontrol strategies, because it does not present a density dependent or a numerical response to increasing densities of *O. perseae* (Aponte & McMurtry, 1997; Hoddle et al., 1999, Lara & Hodle, 2015a and references therein). However, other species of the Phytoseiidae family were very useful in controlling *O. perseae*: for example Aponte et al. (1997) conducted a study in which they monitored populations of the persea mite and its predators, and carried out early augmentative releases of *G. helveolus* and *G. annectens* in several localities of Southern California, where releases of *G. helveolus* has been done in previous years. The results showed that populations of *O. perseae* were reduced after augmentative releases of predaceous mites in plots with previous releases. No differences were found in pest populations between release and control plots in the locality in which no previous augmentative releases were made. The authors ascribed this reduction to the effect of established predators, mainly *G. helveolus*, but also to the combination with *G. annectens*. Another strategy based in augmentative releases was applied using the specific tetranychid predator *N. californicus*. In several studies authors concluded that augmentative releases of *G. helveolus* or *N. californicus*, either alone or in combination, may achieve a good control of *O. perseae* (Hoddle et al., 1999; Kerguelen & Hoddle, 1999), since as stated earlier, both can penetrate inside the *O. perseae* nests and are commercially available.

For a successful biocontrol strategy, it is very important to take into account the best moment to make the releases. Hoddle et al. (1999) determined the timing of releases of six predatory mites, concluding that inoculative releases of *G. helveolus* and/or *N. californicus* may be more effective if releases are made after 25% and before 75% leaf infestation, but further studies are needed.

In Spain, the first avocado producer country in Europe, *O. perseae* was first found in orchards of the Southern Spain production area in 2004 (Vela et al., 2007). As in the USA and other countries where this pest is present, some phytoseiid mite predators have been found as its natural enemies, being *N. californicus* and *E. stipulatus* the two most common species (González-Fernández et al., 2009). In this production area (specifically the coast of Málaga and Granada provinces), field surveys were conducted with the objective of determining the structure of the avocado community by monitoring *O. perseae* and the populations of its predatory enemies during one year (González-Fernández et al., 2009). Phytoseiids and *O. perseae* populations presented two and one peaks of abundance respectively during the year. The identification of the phytoseiid species present in each population peak revealed that the structure of the community varies through time: in spring (April and May), when the pollen amount of several plant species (preferentially olive) was high in the atmosphere, the avocado community was dominated by the omnivore mite *E. stipulatus* that feeds preferentially on pollen grains deposited on the surface of the avocado leaves. In summer, when *O. perseae* reaches its maximum population density, both *E. stipulatus* and *N. californicus* were present, feeding on the persea mite, the shared prey. Based on this result, a biocontrol strategy was conducted with the objective of increasing the simultaneous coexistence of both predator species on time (González-Fernández et al., 2009). In order to do so, field experiments were done by planting maize plants within an avocado orchard, so as to artificially prolong the presence of pollen benefitting *E. stipulatus* populations. Results showed that a source of alternative food such as pollen benefits the phytoseiid abundance, either through the induction of apparent competition between the pollen and the herbivore or by decreasing the potential intraguild predation between predators. Consequently pest density and leaf damage were reduced. The addition of alternative food has also resulted in a better establishment of natural enemies and in an enhanced biocontrol in other systems (Adar et al., 2014; Nomikou et al., 2010, Vandekerkhove & De Clercq, 2010).

In Israel, *O. perseae* was first discovered in 2001. Maoz et al. (2011) carried out an *O. perseae*'s biocontrol strategy at three spatial scales (avocado leaf discs, seedlings and trees) combining the inundative release of the commercial mite *N. californicus* with the conservation of populations of the native mite *E. scutalis* by provisioning it with the pollen of either corn or Rhodes grass (*Chloris gayana* Kunth, Poaceae Family) using different application techniques, depending on the experiment. Their results were similar to those in Spain, showing that the combined release of a specific tetranychid predator and the use of an alternative food for the native omnivore phytoseiid promote a reduction of *O. perseae* populations in avocado orchards.

4. Effects of Abiotic Conditions on Biological Parameters and Interactions between Natural Enemies, and on the Success of BPC Strategies in the Avocado Community

Abiotic factors, such as temperature and relative humidity, are determinant factors influencing many fitness-related parameters (e.g., development, growth, survival, longevity, reproduction, fecundity, etc.) of ectotherm organisms as arthropod pests and their natural enemies. This effect has been evaluated in many studies with phytoseiid mites (Abbassy et al., 2012; Bounfour & McMurtry 1987; Broufas et al., 2007; De Vis et al., 2006; Ferragut et al., 1987; Ferrero et al., 2010; Gadino & Walton, 2012; Kasap, 2009; Raza et al., 2005; Roy et al., 2003; Tsoukanas et al., 2006; Walzer et al., 2007). In the Southern Spain avocado community, the effect of temperature and relative humidity on three fitness-related parameters (oviposition rate, egg hatching and survival) of the two phytoseiid predatory mites *E. stipulatus* and *E. scutalis* was evaluated in fifteen treatments defined by different combinations of temperature (20 – 37ºC) and RH (30 – 75%). The effect of both abiotic factors varied between species. The increase of temperatures affected these three parameters in both species, but, to a greater extent in *E. stipulatus*. However, the relative humidity affected the survival of female, the oviposition rate and the egg hatch of *E. stipulatus*, whereas in *E. scutalis* it only affected the egg hatch (Guzmán et al., in preparation). According to Bounfour & McMurtry (1987) and Ferragut & Escudero (1997), *E. scutalis* seems to be better adapted to harsh environmental conditions (high temperature and low relative humidity) than *E. stipulatus*.

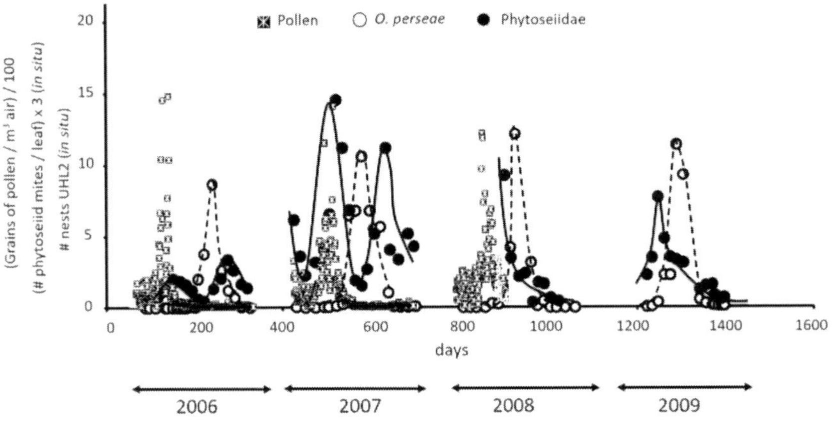

Figure 2. Population dynamics of phytoseiid mites (black dots), pollen grains in the atmosphere (black stars) and the persea mite (white dots), from early spring to late summer, during four consecutive years.

On the other hand, the effect of the environmental conditions on the dynamics of the mite community on avocado was analyzed (Montserrat et al., 2013a). Authors used field data on prey (*O. perseae*) and predators (phytoseiid mites) populations, collected during four years. Data on local temperatures were also included, demonstrating an increased warming tendency in the area of study during the last 28 years. Results showed that in the last two years of the study (2008 and 2009), predators did not respond numerically to the abundance of the pest, unlike what was observed in the first two years (2006 and 2007). The absence of the second phytoseiid peak of abundance in 2008 and 2009 can be observed in Figure 2. It was also found that range of temperatures with positive per capita growth rates was quite larger in prey than in predators, and that the relative humidity explained the growth rate variation in predators, but not in prey. These results could explain the lack of numerical response of the predators during 2008 and 2009, when periods of high temperature and low relative humidity were more frequent, confirming that probably harsh abiotic conditions may disrupt natural biocontrol of pest in agricultural systems.

In the same area, the same authors planned and put into practice a strategy of BPC of the persea mite during two consecutive years (Montserrat et al., 2013b). In this case, the authors chose to increase the populations of *E. stipulatus* through the direct supply of commercial bee pollen as alternative food, sprayed on the tree canopy during the summer (from the beginning of June until the end of September). This addition of alternative food for *E.*

stipulatus was combined with the release of the specialized mite *N. californicus* (1000 – 3500 individual per tree) in the middle of summer, coinciding with the increase of *O. perseae* populations. Results suggested that pollen application promoted a numerical response of the phytoseiids populations, but only during the first half of the summer, so that it failed to achieve a good control of *O. perseae* populations, contradicting the previous work of the same group (González-Fernández et al., 2009). This led the authors to suggest that probably other factors, i.e., the abiotic conditions, were affecting the predator-prey dynamics in the southern Spain avocado community. Thus, additional analyses were included to determine biotic and abiotic effects on the dynamics of this population, concluding that high temperatures during the second half of the summer (these two years were very hot and dry) explained the decline of both predator and prey abundances, and the low rate of pest control.

As we stated before, BPC is in part based on the biotic interactions (predation, competition, etc.) in which pests and predators are involved. As these interactions are climate-dependent, changes in abiotic conditions may interfere with their functioning, potentially affecting BPC strategies. Thus, to reach a successful BPC, it is also necessary to determine how biotic interactions may vary in function with temperature and relative humidity, among other factors. In the avocado system, Guzmán et al. (2016a) analyzed the effect of these abiotic conditions on the strength and direction of the intraguild predation that may occur between the two phytoseiid mites *E. stipulatus* and *E. scutalis*, predators of the avocado persea mite. Intraguild predation may happen when both species (the IG-predator and the IG-prey) interact through both predation and competition. The existence of this type of interaction is useful to be determined, as in some cases, intraguild predation can cause disruption of BPC (Rosenheim et al., 1995), although in the field, its negative impact can be attenuated by different mechanisms, such as spatial heterogeneity (Janssen et al., 2006, 2007) or anti-predator behavior (Magalhães et al., 2005). Results showed that in avocado *E. stipulatus* and *E. scutalis* did not engage on intraguild predation, as both species preferentially feed on the shared resource (pollen) instead of feeding on the IG-prey (the other specie). Moreover, IG-prey mortality was not affected by abiotic conditions in any of the two species. Based on the absence of intraguild predation, authors concluded that the dynamics of these two species in the avocado fields is mainly determined by exploitative resource competition. This conclusion was also confirmed in another study: the carrying capacity (k) and interspecific competition term (α) were determined for each predatory mite species (*E. stipulatus* and *E. scutalis)* through population and community laboratory experiments, to be able to predict the dominance of species under two abiotic conditions, “mild” (25ºC and 75% RH) and “hot-dry” (30ºC and 50% RH) (Guzmán et al., 2016b). Field sampling were also carried out in areas under both conditions. Results from model simulations and from field samplings showed that abiotic conditions profoundly affected the competitive ability of both species, resulting in changes in the structure of the community depending on the temperature. Under mild conditions, both species seemed to exploit the resource with similar efficiency (Figure 3, left panel). However, simulations revealed that the outcome of the interaction was determined by priority effects, which explained why *E. stipulatus* (i.e., the local and resident species) was the only one present in the coastal orchards, where abiotic summer conditions are relatively mild. Under hot-dry conditions, *E. scutalis* was the superior competitor (Figure 3, right panel), probably because it is better adapted to such conditions than *E. stipulatus*, which explained why it is the only species found in inland avocado orchards, where summer abiotic conditions are more extreme.

These works showed the relevance of abiotic conditions when planning biocontrol strategies in the present climate change scenarios.

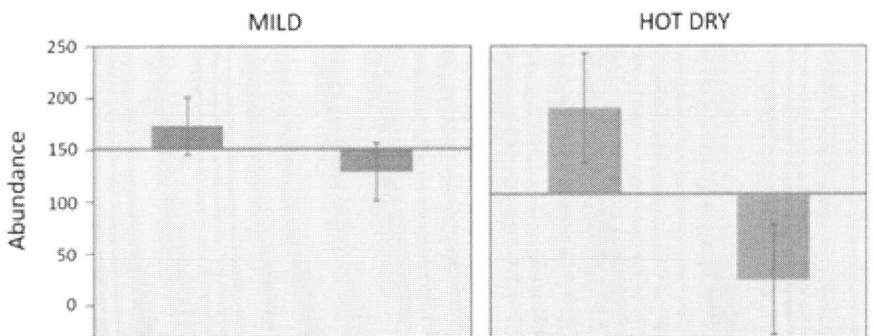

Figure 3. Equilibrium population levels of *Euseius scutalis* and *E. stipulatus* at the two abiotic conditions (mild and hot-dry). Horizontal lines represent the average global population; columns are deviation contrasts that depict significant differences between the average global population and population levels of each species, if the standard deviations do not cross the horizontal line.

Conclusion

O. perseae is an avocado pest detected in several producer countries where different avocado-inhabiting species of phytoseiid predators have the potential to exert a natural control of its populations. Usually, BPC strategies were necessary to achieve a good control of this tetranychid pest. In avocado orchards, effective strategies for *O. perseae* control have consisted in increasing the predator populations, either by the addition of a high number of commercial individuals (augmentative BPC, with *N. californicus* and *G. helveolus*), or by supplementing alternative food as pollen (conservation BPC), or even through the combination of both methods. Nevertheless, it has recently been proved that abiotic conditions (mainly temperatures and relative humidity) may be a direct cause for the disruption of *O. perseae*'s biological control. Therefore, given the inevitability of climate change, has become necessary to include the effects of abiotic conditions on biological parameters of pests and natural enemies, and on the strength of their interactions, if we are to design effective biocontrol strategies in the future.

References

Abbassy, M. R., Hendy, H. H., Mowafi, M. H. & Nawar, M. A. (2012). Biology of *Euseius scutalis* (Acari: Phytoseiidae) on *Tetranychus urticae* and *Panonychus ulmi* (Acari: Tetranychidae) at different temperatures. *Acarines*, *6*, 15-19.

Adar, E., Inbar, M., Gal, S., Gan-Mor, S. & Palevsky, E. (2014). Pollen on-twine for food provisioning and oviposition of predatory mites in protected crops. *BioControl*, *59*, 307-317.

Aponte, O. & McMurtry, J. A. (1997). Damage on 'Hass' avocado leaves, webbing and nesting behaviour of *Oligonychus perseae* (Acari: Tetranychidae). *Experimental and Applied Acarology*, *21*, 265–272.

Aponte, O., Hoddle, M., Heraty, J., Personnel, C., Johnson, H. & Pinckard, T. (1997). Biological control of pests on avocado in California. *California Avocado Research Symposium*, *1997*, 13-18.

Bender, G. S. (1993). A new mite problem in avocados. *California Avocado Society Yearbook*, *77*, 73-77.

Beveridge, O. S., Humphries, S. & Petchey, O. L. (2010). The interacting effects of temperature and food chain length on trophic abundance and ecosystem function. *Journal of Animal Ecology*, *79*, 693–700.

Borges, P. A. V, Aguiar, A. M. F, Boieiro, M., Carles-Tolrá, M. & Serrano, A. R. M. (2008). List of arthropods (Arthropoda) of madeira and Selvagens archipelagos. In: Borges, P. A. V., Aguiar, A. M. F., Carvalho, P., Jardim, R., Melo, I., Oliveira, P., Sérgio, C., Serrano, A. R. M., Vieira, P. (Eds). A list of the terrestrial fungi, flora and fauna of Madeira and Selvagens archipelagos. Direcção Regional do Ambiente da Madeira and Universidade dos Açores, Funchal and Angra do Heroísmo, Portugal; p. 440.

Bounfour, M. & McMurtry J. A. (1987). Biology and ecology of *Euseius scutalis* (Athias-Henriot) (Acarina: Phytoseiidae). *Hilgardia*, *55*, 1-23.

Broufas, G. D., Pappas, M. L. & Koveos, D. S. (2007). Development, survival, and reproduction of the predatory mite *Kampimodromus aberrans* (Acari: Phytoseiidae) at different constant temperatures. *Environmental Entomology*, *36*, 657-665.

Cagan, L., Tancik, J. & Hassan, S. (1998). Natural parasitism of the European corn borer eggs *Ostrinia nubilalis* Hbn. (Lep., Pyralidae) by *Trichogramma* in Slovakia—need for field releases of the natural enemy. *Journal of Applied Entomology*, *122*, 315-318.

Cock, M. J., Biesmeijer, J. C., Cannon, R. J., Gerard, P. J., Gillespie, D., Jimenez, J. J., Lavelle, P. M. & Raina, S. K. (2013). The implications of climate change for positive contributions of invertebrates to world agriculture. *CAB Reviews*, *8*, 1–48.

Coll, M. & Hughes, L. (2008). Effects of elevated CO_2 on an insect omnivore: a test for nutritional effects mediated by host plants and prey. *Agriculture, Ecosystems & Environment*, *123*, 271-279.

De Vis, R. M. J., de Moraes, G. J. & Bellini, M. R. (2006). Effect of air humidity on the egg viability of predatory mites (Acari: Phytoseiidae, Stigmaeidae) common on rubber trees in Brazil. *Experimental & Applied Acarology*, *38*, 25-32.

Dunson, W. A. & Travis, J. (1991) The role of abiotic factors in community organization. *The American Naturalist*, *138*, 1067–1091.

English-Loeb, G. M. (1990). Plant drought stress and outbreaks of spider mites: a field test. *Ecology*, *71*, 1401-1411.

FAOSTAT. Food and Agriculture Organization of the United Nations. Statistics Division. Data available at: http://faostat3.fao.org.

Ferragut, F. J. & Escudero, A. (1997). Taxonomía y distribución de los ácaros depredadores del género *Euseius* Wainstein 1962, en España (Acari: Phytoseiidae). *Boletín de Sanidad Vegetal Plagas*, *23*, 227-235. [Taxonomy and distribution of predatory mites of the genus *Euseius* Wainstein 1962, in Spain (Acari: Phytoseiidae). *Plant Health Bulletin Pests, 23*, 227-235.]

Ferragut, F., Garcia-Mari, F., Costa-Comelles, J. & Laborda, R. (1987). Influence of food and temperature on development and oviposition of *Euseius stipulatus* and *Typhlodromus phialatus* (Acari: Phytoseiidae). *Experimental and Applied Acarology*, *3*, 317-329.

Ferrero, M., Gigot, C., Tixier, M. S., van Houten, Y. M. & Kreiter, S. (2010). Egg hatching response to a range of air humidities for six species of predatory mites. *Entomologia Experimentalis et Applicata. 135*, 237–244.

Gadino, A. N. & Walton, V. M. (2012). Temperature-related development and population parameters for *Typhlodromus pyri* (Acari: Phytoseiidae) found in Oregon vineyards. *Experimental and Applied Acarology*, *58*, 1-10.

Gilman, S. E., Urban, M. C., Tewksbury, J., Gilchrist, G. W. & Holt, R. D. (2010). A framework for community interactions under climate change. *Trends in Ecology and Evolution*, *25*, 325–331.

González-Fernández, J. J., de la Peña, F., Hormaza, J. I., Boyero, J. R., Vela, J. M., Wong, E., Trigo, M. M. & Montserrat, M. (2009). Alternative food improves the combined effect of an omnivore and a predator on biological pest control. A case study in avocado orchards. *Bulletin of Entomological Research*, *99*, 433-444.

Guzmán, C., Sahún, R. M. & Montserrat, M. (2016a). Intraguild predation between phytoseiid mite species might not be so common. *Experimental and Applied Acarology*, *68*, 441-453.

Guzmán, C., Aguilar-Fenollosa, E., Sahún, R. M., Boyero, J. R., Vela, J. M., Wong, E., Jaques J. A. & Montserrat, M. (2016b). Temperature-specific competition in predatory mites: Implications for biological pest control in a changing climate. *Agriculture, Ecosystems & Environment*, *216*, 89-97.

Hoddle, M. S., Aponte, O., Kerguelen, V. & Heraty, J. (1999). Biological control of *Oligonychus perseae* (Acari: Tetranychidae) on avocado: I. Evaluating release timings, recovery and efficacy of six commercially available phytoseiids. *International Journal of Acarology*, *25*, 211-219.

Hoddle, M. S. & Morse, J. G. (2013). The persea mite invasion into California: history, biology, management and current status. *California Avocado Society Yearbook*, *95*, 106-136.

IPCC (2014). Climate Change 2014: Synthesis Report. Contribution of Working Groups I, II and III to the Fifth Assessment. In Core Writing Team, R. K. Pachauri & L. Meyer (Eds.), *Report of the Intergovernmental Panel on Climate Change*. 151 pp. Geneva, Switzerland: IPCC.

Janssen, A., Montserrat, M., HilleRisLambers, R., de Roos, A. M., Pallini, A. & Sabelis, M. W. (2006). Intraguild predation usually does not disrupt biological control. In J. Brodeur & G. Boivin (Eds.), *Progress in Biological Control: Trophic and Guild Interactions in Biological Control*, Volume *3* (pp. 21–44), The Netherlands: Springer.

Janssen, A., Sabelis, M. W., Magalhães, S., Montserrat, M. & van der Hammen, T. (2007). Habitat structure affects intraguild predation. *Ecology*, *88*, 2713–2719.

Kasap, I. (2009). Influence of temperature on life table parameters of the predaceous mite *Euseius finlandicus* (Oudemans) (Acari: Phytoseiidae). *Turkish Journal of Agriculture and Forestry*, *33*, 29-36.

Kerguelen, V. & Hoddle, M. S. (1999). Biological control of *Oligonychus perseae* (Acari: Tetranychidae) on avocado: II. Evaluating the efficacy of *Galendromus helveolus* and *Neoseiulus californicus* (Acari: Phytoseiidae). *International Journal of Acarology*, *25*, 221-229.

Lara, J. R. & Hoddle, M. S. (2013). Sampling guidelines for persea mite in California avocado orchards. *California Avocado Society Yearbook*, *95*, 151–173.

Lara, J. R. & Hoddle, M. S. (2015a). Past, Present, and Future: Biological Control of Spider Mites on California-Grown Avocados. In Prospects for Biological Control of Plant Feeding Mites and Other Harmful Organisms (pp. 271-308). Springer International Publishing.

Lara, J. R. & Hoddle, M. S. (2015b). Comparison and field validation of binomial sampling plans for *Oligonychus perseae* (Acari: Tetranychidae) on Hass Avocado in Southern California. *Journal of Economic Entomology*, *108*, 2074-2089.

Lara, J. R., Saremi, N. T., Castillo, M. J. & Hoddle, M. S. (2016). Sampling method evaluation and empirical model fitting for count data to estimate densities of *Oligonychus perseae* on 'Hass' avocado leaves in southern California. *Experimental and Applied Acarology*, *68*, 455-475.

Magalhães, S., Tudorache, C., Montserrat, M., van Maanen, R., Sabelis, M. W. & Janssen, A. (2005). Diet of intraguild predators affects antipredator behavior in intraguild prey. *Behavioral Ecology*, *16*, 364–370.

Maoz, Y., Gal, S., Argov, Y., Coll, M. & Palevsky, E. (2011). Biocontrol of persea mite, *Oligonychus perseae* with an exotic spider mite predator and an indigenous pollen feeder. *Biological Control*, *59*, 147-157.

McMurtry, J. A. & Croft, B. A. (1997). Life-styles of phytoseiid mites and their roles in biological control. *Annual Review of Entomology*, *42*, 291-321.

Memmott, J., Craze, P. G., Waser, N. M. & Price, M. V. (2007). Global warming and the disruption of plant–pollinator interactions. *Ecology Letters*, *10*, 710-717.

Montserrat, M., de la Peña, F., Hormaza, J. I. & González-Fernández, J. J. (2008). How do *Neoseiulus californicus* (Acari: Phytoseiidae) females penetrate densely webbed spider mite nests? *Experimental and Applied Acarology*, *44*, 101-106.

Montserrat, M., Sahún, R. M. & Guzmán, C. (2013a). Can climate change jeopardize predator control of invasive herbivore species? A case study in avocado agro-ecosystems in Spain. *Experimental and Applied Acarology*, *59*, 27-42.

Montserrat, M., Guzmán, C., Sahún, R. M., Belda, J. E. & Hormaza, J. I. (2013b). Pollen supply promotes, but high temperatures demote, predatory mite abundance in avocado orchards. *Agriculture, Ecosystems & Environment*, *164*, 155-161.

Nomikou, M., Sabelis, M. W. & Janssen, A. (2010). Pollen subsidies promote whitefly control through the numerical response of predatory mites. *BioControl*, *55*, 253–260.

Preisser, E. L. & Strong, D. R. (2004). Climate affects predator control of an herbivore outbreak. *The American Naturalist*, *163*, 754-762.

Raza, A. B. M., Afzal, M. & Bashir, M. H. (2005). Biology of *Euseius septicus* Chaudhari (Acari: Phytoseiidae) preying on two spotted spider mite, *Tetranychus urticae* (Acari: Tetranychidae) at different temperatures. *Pakistan Entomologist*, *27*, 85-88.

Rosenheim, J. A., Kaya, H. K., Ehler, L. E., Marois, J. J. & Jaffee, B. A. (1995). Intraguild predation among biological-control agents: theory and evidence. *Biological Control*, *5*, 303-335.

Roy, M., Brodeur, J. & Cloutier, C. (2003). Effect of temperature on intrinsic rates of natural increase (r_m) of a coccinellid and its spider mite prey. *BioControl*, *48*, 57-72.

Salas, L. (1978). Algunas notas sobre las arañitas rojas (Tetranychidae: Acari) halladas en Costa Rica. *Agronomía Costarricense*, *2*, 47–59. [Some Notes on the spider mites (Tetranychidae: Acari) found in Costa Rica. *Costa Rican Agronomy*, *2*, 47-59]

Smith, N. J. H., Williams, J. T., Plucknett, D. L. & Talbot, J. P. (1992). Tropical Forest and Their Crops. Ithaca: Cornell University Press.

Snyder, W. E. & Evans, E. W. (2006). Ecological effects of invasive arthropod generalist predators. *Annual Review of Ecology, Evolution and Systematics*, *37*, 95-122.

Thomson, L. J., Macfadyen, S. & Hoffmann, A. A. (2010). Predicting the effects of climate change on natural enemies of agricultural pests. *Biological Control*, *52*, 296-306.

Tsoukanas, V. I., Papadopoulos, G. D., Fantinou, A. A. & Papadoulis, G. T. (2006). Temperature-dependent development and life table of *Iphiseius degenerans* (Acari: Phytoseiidae). *Environmental Entomology*, *35*, 212-218.

Tylianakis, J. M., Didham, R. K., Bascompte, J. & Wardle, D. A. (2008). Global change and species interactions in terrestrial ecosystems. *Ecology Letters*, *11*, 1351-1363.

Vandekerkhove, B. & De Clercq, P. (2010). Pollen as an alternative or supplementary food for the mirid predator *Macrolophus pygmaeus*. *Biological Control*, *53*, 238-242.

Vela, J. M., González-Fernández, J., Wong, E., Montserrat, M., Farré, J. M. & Boyero, J. R. (2007). El ácaro del aguacate (*Oligonychus perseae*): Estado actual del problema e investigación en Andalucía. *Agrícola Vergel*, *306*, 301–308. [The avocado mite (*Oligonychus perseae*): Current status of the problem and research in Andalusia. *Agrícola Vergel*, *306*, 301-308.]

Voigt, W., Perner, J., Davis, A. J., Eggers, T., Schumacher, J., Bahrmann, R., Fabian, B., Heinrich, W., Kohler, G., Lichter, D., Marstaller, R. & Sander, F. W. (2003). Trophic levels are differentially sensitive to climate. *Ecology*, *84*, 2444–2453.

Walzer, A., Castagnoli, M., Simoni, S., Liguori, M., Palevsky, E. & Schausberger, P. (2007). Intraspecific variation in humidity susceptibility of the predatory mite *Neoseiulus californicus:* survival, development and reproduction. *Biological Control*, *41*, 42–52.

Williams, L. O. (1977). The avocados, a synopsis of the genus *Persea*, subg. *Persea Economic Botany*, *31*, 315-320.

Zappalà, L., Kreiter, S., Russo, A., Tropea Garzia, G. & Auger, P. (2015). First record of the persea mite *Oligonychus perseae* (Acari: Tetranychidae) in Italy with a review of the literature. *International Journal of Acarology*, *41*, 97-99.

Biographical Sketches

Celeste Guzmán

Estación Experimental de Zonas Áridas (EEZA).
Spanish National Research Council (CSIC)

Education: PhD in Biology (University of Málaga, Spain). Master in Celular and Molecular Biology (University of Málaga, Spain).

Research and Professional Experience: Community ecology, population dynamics, biological pest control

Professional Appointments: PhD scholarship from the Andalusian Government (Becas de Excelencia de la Junta de Andalucía) (2009-2013); Postdoctoral grants from the Andalusian Government (2014-2015 and 2016-2018)

Publications Last Three Years:

Guzmán C, Sahún RM, Montserrat M (2016) Intraguild predation between phytoseiid mite species might not be so common. *Experimental and Applied Acarology,* 68: 441-453.

Guzmán C, Aguilar-Fenollosa E, Sahún RM, Boyero JR, Vela JV, Wong E, Jaques JA, Montserrat M. (2016). Temperature-specific competition in predatory mites: implications for biological pest control in a changing climate. *Agriculture, Ecosystems & Environment,* 216: 89-97.

Montserrat, M., Guzmán C., Sahún R.M., Belda J.E., Hormaza J.I. (2013) Pollen supply promotes, but high temperature demote, predatory mite abundance in avocado orchards. *Agriculture Ecosystems & Environment*, 164: 155-161.

Montserrat M., Sahún R.M., Guzmán C. (2013) Can climate change jeopardize predator control of invasive herbivore species? A case study in avocado agro-ecosystems in Spain. *Experimental and Applied Acarology*, 59: 27-42.

C. Guzmán, E. Aguilar-Fenollosa, I. Torres-Campos, R.M Sahún, J.A. Jacas, M. Montserrat (2013). Environmental conditions affect competitive and predatory interactions among natural enemies. Implications for pest biological control in a changing climate. Peter G. Mason, David R. Gillespie and Charles Vincent (Eds.). Proceedings of the 4th International Symposium on Biological Control of Arthropods. Pucón, Chile, pp. 4-8.

Marta Montserrat Larrosa

Instituto de Hortofruticultura Subtropical y Mediterránea "La Mayora" (IHSM-UMA-CSIC)

Education: PhD in Agricultural Sciences (LLeida University, Spain); PhD in Ecology (University of Amsterdam, The Netherlands)

Research and Professional Experience: Food web engineering, community ecology, population dynamics, conservation biological pest control

Professional Appointments: Tenured scientist at the Spanish National Research Council (CSIC); Associate Editor of the journals Biocontrol and International Journal of Pest Management

Publications Last Three Years:

Guzmán, C., Sahún, R.M., Montserrat, M. (2016) Intraguild predation between phytoseiid mite species might not be so common. *Experimental and Applied Acarology*, 68: 441-453.

Guzmán, C., Aguilar-Fenollosa, E., Sahún, R.M., Boyero, J.R., Vela, J.V., Wong, E., Jaques, J.A., Montserrat, M. (2016). Temperature-specific competition in predatory mites: implications for biological pest control in a changing climate. *Agriculture, Ecosystems & Environment,* 216: 89-97.

Moya-Laraño, J., Bilbao-Castro, J.R., Barrionuevo, G., Ruiz-Lupión, D., Casado, L.G., Montserrat, M., Melian, C., Magalhaes, S. (2014) Eco-evolutionary spatial dynamics: rapid evolution and isolation explain food web persistence. *Adv. Ecol. Res.* 50:75-143.

Montserrat, M., Guzmán C., Sahún R.M., Belda J.E., Hormaza J.I. (2013) Pollen supply promotes, but high temperature demote, predatory mite abundance in avocado orchards. *Agriculture, Ecosystems & Environment*, 164: 155-161.

Montserrat, M., Sahún, R.M., Guzmán, C. (2013) Can climate change jeopardize predator control of invasive herbivore species? A case study in avocado agro-ecosystems in Spain. *Experimental and Applied Acarol*ogy, 59: 27-42.

Montserrat, M., Requena, J.M., Valladares, F., Rodríguez, E., González, M., De Mas, E., Moya-Laraño, J. (2013). Sobre primaveras silenciosas y mundos verdes: ¿Tan utópica es una agricultura sin plaguicidas? Article in eldiario.es. http://www.eldiario.es/cienciacritica/primaveras-silenciosas-utopica-agricultura-plaguicidas_6_191490872.html.

Guzmán, C., Aguilar-Fenollosa, E., Torres-Campos, I., Sahún, R.M., Jacas, J.A., Montserrat, M. (2013). Environmental conditions affect competitive and predatory interactions among natural enemies. Implications for pest biological control in a changing climate. Peter G. Mason, David R. Gillespie and Charles Vincent (Eds.). Proceedings of the 4th International Symposium on Biological Control of Arthropods. Pucón, Chile, pp. 4-8.

In: Natural Enemies
Editor: Sónia A.P. Santos

ISBN: 978-1-63485-921-9

Chapter 3

BIODIVERSITY OF SPIDERS IN AGROECOSYSTEMS: COMMUNITY STRUCTURE, CONSERVATION AND ROLES AS BIOLOGICAL CONTROL AGENTS

Jacinto Benhadi-Marín[1,2], José A. Pereira[1], Albino Bento[1], José Paulo Sousa[2] and Sónia A. P. Santos[1,3,*]

[1]Mountain Research Center (CIMO), School of Agriculture, Polytechnic Institute of Bragança, Bragança, Portugal
[2]Centre for Functional Ecology, Department of Life Sciences, University of Coimbra, Coimbra, Portugal
[3]Barreiro School of Technology, Polytechnic Institute of Setúbal, Lavradio, Portugal

ABSTRACT

Spiders are one of the most abundant and diverse groups of predators in agroecosystems. They hunt mainly insects in all of their instars and, for this reason, together with their ubiquity and high diversity of hunting

* Corresponding author: Mountain Research Center (CIMO), School of Agriculture, Polytechnic Institute of Bragança, Campus of Santa Apolónia, 5300-253 Bragança, Portugal. Email: saps@ipb.pt.

strategies, they can be successful biological control agents of arthropods pests. Although, agricultural management practices such as the use of pesticides, tillage or fertilization, landscape simplification and intensive farming may alter the diversity of spiders and their effectiveness in the agroecosystem. On the other hand, within the agroecosystem, ecological infrastructures as vegetation strips, small bushes and stonewalls can provide shelter and alternative prey that can be exploited by a range of functional groups of spiders. Aspects such as the conservation of a heterogeneous agricultural landscape as well as the maintenance of ecological infrastructures can be seen as part of a strategy to achieve sustainability and promote conservation biological control. This chapter provides an overview of the effects of the most common agricultural practices on the biodiversity and life-history parameters of spiders using complementary approaches, i.e., traditional taxonomy and functional groups, and how these pressures may change the ecological services provided by spiders. A case study regarding the community of spiders associated with the olive grove agroecosystem will be described.

Keywords: Araneae, predator, landscape, crops, traits, bioinsecticides

1. INTRODUCTION

Arachnids already existed in the Carboniferous, between 280 and 350 million years ago (Hubert, 1979). Currently, 45.881 species of spiders (order Araneae) distributed in 3.981 genera and 114 families are described worldwide (World Spider Catalog, 2016) and, together with Amblypygi, Holothyrida, Ixodida, Mesostigmata, Opilioacarida, Opiliones, Palpigradi, Pseudoscorpiones, Ricinulei, Scarcoptiformes, Schizomida, Scorpiones, Solifugae, Thelyphonida and Trombidiformes, they are one of the 16 arthropod orders that form the class Arachnida (Zhang, 2011). Araneae order is divided into two major suborders, Mesothelae and Opisthothelae. The suborder Mesothelae includes primitive spider families characterized by the presence of segments on the opisthosoma. The suborder Opisthothelae comprises the infraorders Mygalomorphae (commonly known as tarantulas) and Araneomorphae (the vast majority of described species including jumping, wolf, orb-weaving spiders, etc.) (Bond et al., 2014; Dunlop & Penney, 2011). The infraorder Mygalomorphae includes spiders with a morphological character that separate them from the Araneomorphae, the opening direction of the chelicerae, which are almost parallel to the cephalocaudal axis in Mygalomorphae and crossed in Araneomorphae (Figure 1) (Barrientos, 1988).

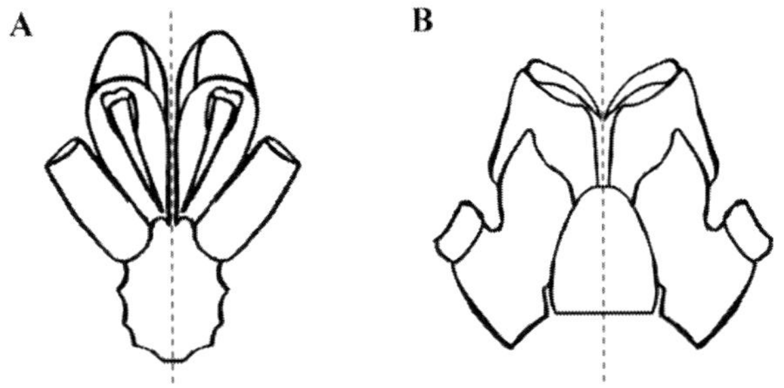

Figure 1. Opening direction of the chelicerae of spiders; (A) almost parallel to the cephalocaudal axis in Mygalomorphae and (B) crossed in Araneomorphae. Dashed lines follow the cephalocaudal axis. Illustration by J. Benhadi-Marín.

Spiders represent one of the most diverse groups of invertebrates and are distributed worldwide, virtually adapted to any type of habitat where other arthropods, the basic component of their diet, are present. They exist in the northernmost Arctic islands, in the hottest and driest deserts, in high altitudes, in deep caves, in the intertidal zone of coastal areas, ponds and marshes, in arid moors, sand dunes, and floodplains. Also, spiders have invaded the aquatic environment and successfully compete with other predatory invertebrate communities in shallow water systems (Turnbull, 1973). The Liphistiomorpha group is restricted to the Southeast of Asia, China and Japan (World Spider Catalog, 2016). Theraphosomorpha is characteristic of tropical and subtropical areas, however, it extends to colder areas both to the north and south and considering their extreme diversity can be expected that Araneomorpha occupy all kinds of habitats (Turnbull, 1973).

Spiders are generalist predators, which hunt mainly insects of a wide range of sizes and are considered as potential natural enemies of pests. They also exploit non-prey foods such as pollen and nectar (e.g., Nyffeler et al., 2016) and there is an increasing interest in understanding how alternative food sources affect the community of spiders in order to implement ecological infrastructures (e.g., wildflower strips or ground covers) in the agroecosystem that will provide such resources aiming at maintaining and enhancing the abundance of spiders within crop areas. In general, spider-prey interactions have been documented through direct observations of spiders hunting and feeding on prey, morphological measurements, serological and molecular gut-content analysis and studies about the feeding behavior and functional responses of spiders fed on different types of prey (e.g., Greenstone & Shufran, 2003; Ghavami, 2008; Harwood et al., 2009; Monzó et al., 2009; Morris et al., 1999; Nakamura & Nakamura, 1977; Nentwig & Wissel, 1986;

Nyffeler et al., 2016; Oelbermann & Scheu, 2009; Samu & Bíró, 1993; Sunderland et al., 1986). However, crop management such as spraying of pesticides and tillage, a common practice used to control weeds and conserve soil water (Martins et al., 2005) can influence the efficiency of spiders as biological control agents.

This chapter aims at providing an overview of the effects of the most common agricultural management practices on the biodiversity and life-history parameters of spiders using two complementary approaches: (1) how discrete species (i.e., the traditional taxonomy point of view) and (2) functional groups (focused on different groups of species using the same resource in a similar way) respond to environmental changes caused by market pressures within the agricultural landscape at different habitat scales. Also, the emergent interest on the use of spider-venom based products as bioinsecticides is commented and a case study on the diversity of spiders and performance in olive groves is provided.

2. Morphology of Spiders

Spiders have the body divided in two parts, prosoma or carapace (anterior) and opisthosoma or abdomen (posterior) (Figure 2A). Prosoma and opisthosoma are joined by a pedicel corresponding to the first segment of the abdomen (Figure 2B) (Barrientos, 1988; Brusca & Brusca, 2005; Foelix, 1996; Nieto & Mier, 1985).

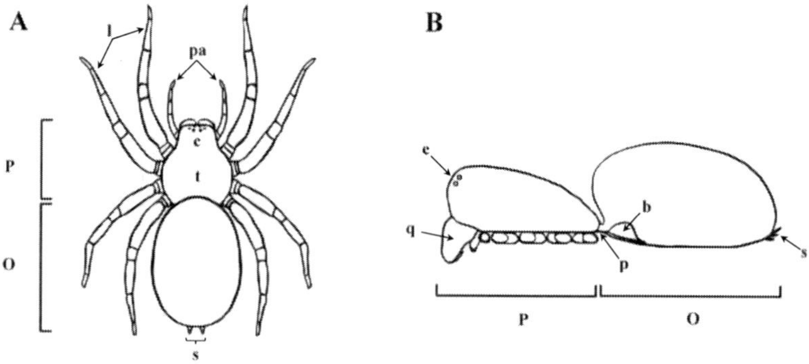

Figure 2. External anatomy of a spider: (A) dorsal view and (B) lateral view. P: prosoma; O: Opisthosoma; l: legs; pa: palps; c: cephalic area; t: thoracic area; s: spinnerets; e: eyes; q: chelicerae; b: book lungs opening and p: pedicel. Illustration by J. Benhadi-Marín.

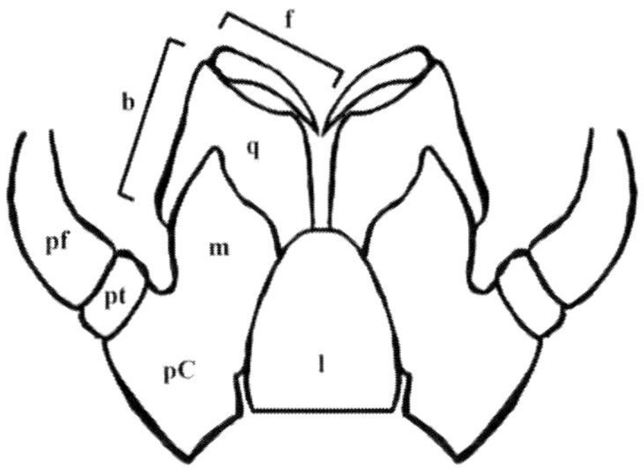

Figure 3. Cephalic region of a spider (simplified) in ventral view. f: cheliceral fang; b: basal cheliceral segment; q: chelicerae; m: maxilla; pC: palpal coxa; pt: palpal trochanter; pf: palpal femur and l: labium. Illustration by J. Benhadi-Marín.

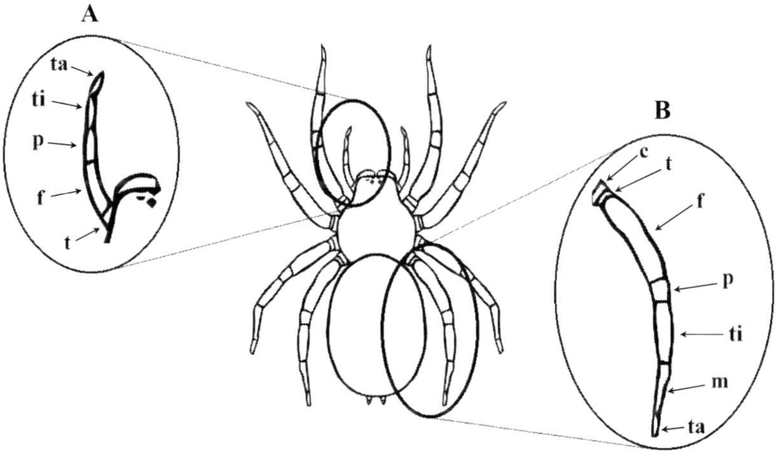

Figure 4. Detail of a spider (A) palp and (B) leg segmentation. c: coxa (not shown for the palp); t: trochanter; f: femur; p: patella; ti: tibia; m: metatarsus; ta: tarsus. Illustration by J. Benhadi-Marín.

Within the prosoma two regions are distinguished, an anterior (cephalic) containing structures related to the senses and food and a posterior (thoracic) related with locomotion and housing the central nervous system (Foelix, 1996). Most of the external appendages on the spider are attached to the cephalothorax, including eight legs, six to eight eyes (exceptionally none in some troglophile spiders), chelicerae, other mouthparts, and palps. The first pair of appendages on the prosoma is formed by the chelicerae, each one divided into a basal segment and a fang (Figure 3). On the apical part of each fang ends the duct that connects the fang with the venom glands, which are located on both sides of the cephalic region of the prosoma.

The second pair of appendages corresponds to the palps, made up by six segments (i.e., coxa, trochanter, femur, patella, tibia and tarsus) (Figure 4A). Legs have the same parts plus the metatarsus between the tibia and the tarsus (Figure 4B). The esophagus, as well as the stomach is also located in this division (Barrientos, 1988; Brusca & Brusca, 2005; Mier, 1985; Nieto & Foelix, 1996).

On the anterior-ventral part of the opisthosoma are the genital openings and the book-lung spiracles openings under the epigastric furrow. The heart, testes or ovaries, midgut, hindgut and spinning glands are also housed in the opisthosoma. Normally three pairs of spinnerets are present on the body posterior end. Finally, the hairs occurring on the body of the spider are sensory organs that response to mechanical or chemical stimulus. Most of these sensilla are movable, articulated setae, bristles and trichobothria, which are mechanoreceptors located on the legs (Barrientos, 1988; Foelix, 1996).

3. Development and Growth of Spiders

As arthropods, the development of spiders has different stages from the egg (Figure 5) to the adult stage. At the end of each growing period, the entire cuticle is substituted by a new one through molting (Figure 6). In spiders, the number of molts varies among groups. European species (depending on the size) molt between five and ten times before reaching the adult stage (Roberts, 1996). During the molting period, the spider is more susceptible to die due to dehydration, failures along the molting process, or predation. Once individuals reach the last juvenile stage, they present the appearance of an adult, but still lack reproductive structures (Foelix, 1996; Turnbull, 1973).

In general, the life cycle of a spider is completed during one year, although it can vary from a few months to several years, depending on the species, the environmental conditions and the occupation of the habitat. Depending on when spiders reach sexual maturity, two types of cycles can be observed: (1) those spiders that reach sexual maturity in spring and reproduce during the summer, spiderlings will be in that stage during the winter, and (2) those spiders that mature during the fall and overwinter as adults (Cárdenas, 2008).

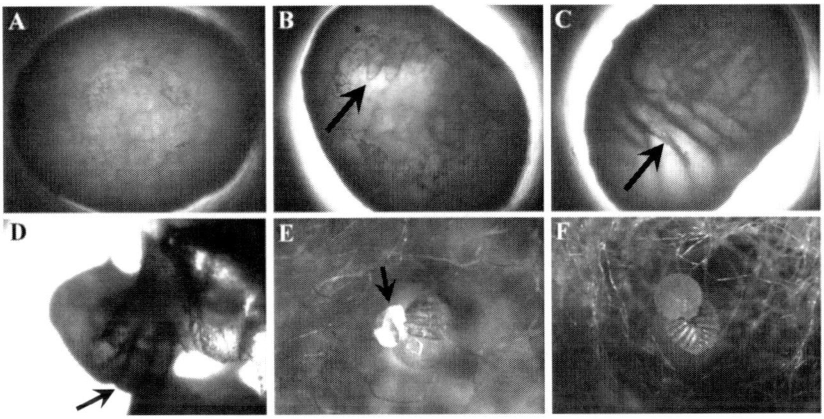

Figure 5. Development of an egg of spider. (A) Early egg stage; (B) Tissues and organs development; (C) Prosoma and opisthosoma regions can be distinguished; (D) Spiderling early hatching stage; (E) Final hatching stage of the spiderling; (F) Spiderling completely hatched. In B, C and D, the arrows indicate different developmental stages of the leg. In E the arrow indicates the remains of the egg shell. Photos by: J. Benhadi-Marín.

Figure 6. Molting of a tarantula (Mygalomorphae). After removing the whole old cuticle (A), the spider stands still (B) until the new cuticle hardens (C) and the spider continues with its activity (D). The arrow in A indicates the old carapace cuticle. Photos by J. Benhadi-Marín.

4. Reproduction of Spiders

Spiders reproduce sexually, however, males and females can present different levels of sexual dimorphism, very pronounced in the case of Thomisidae, whose males are much smaller and graceful than females (Morse, 2007) and an extreme case can be observed in *Nephila* Leach, 1815 (Nephilidae) species (e.g., Kuntner & Coddington, 2009). Males of spiders reach maturity after the last molt, when palps appear completely modified (Figure 7) (Foelix, 1996). Mature male palp is a kind of hypodermic noodle that will be used to transfer, by injection, the sperm into the genital ducts of females.

In general, females show relatively sedentary habits and males are responsible for finding females attracted by sex pheromones (Foelix, 1996; Turnbull, 1973). This risky action can end in several ways, (1) the male dies before finding a female, (2) the male finds a female but the latter is not mature or (3) the female does not find the male attractive. In the two last cases, the male may be seen as a prey by the female and can be attacked and consumed (Foelix, 1996; Turnbull, 1973).

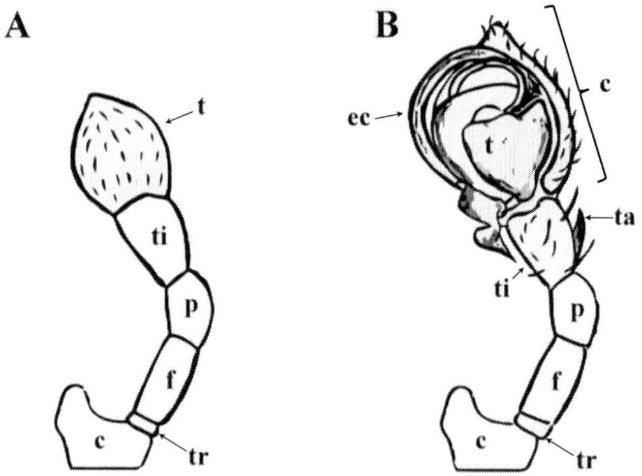

Figure 7. Morphological differences between an immature male spider palp (A) and a mature palp (B). t: tarsus; ti: tibia; p: patella; f: femur; tr: trochanter; c: coxa; c: cimbium; ce: conductor and embolus (the later protected by the former); ta: tibial apophysis. Illustration by J. Benhadi-Marín.

Males of a great number of species can have complex courtships in order to minimize the risk of being eaten. Courtships seem to fulfill three functions: (1) to allow the male to assess if female is ready for mating from a safe distance, (2) to stimulate the female and put it into a receptive state and (3) to allow the female to discriminate different males (Platnick, 1971; Turnbull, 1973). If the male is accepted by the female, then mating occurs (Figure 8).

The level of egg care varies between groups but, in general, the female lays its eggs in a sac-like structures made of silk (Figure 9A). These cocoons can contain from one to hundreds of eggs and can be more or less elaborated using different types of silk, coated substrate fragments (Figure 9B), and in some cases, transported under the abdomen fixed by the spinnerets (e.g., Lycosidae). During the life of a female, the first cocoon usually contains the highest number of eggs that gradually decrease in the subsequent ones (Turnbull, 1973).

Figure 8. Mating in *Haplodrassus severus* (Gnaphosidae). F: female and M: male. Photo by J. Benhadi-Marín.

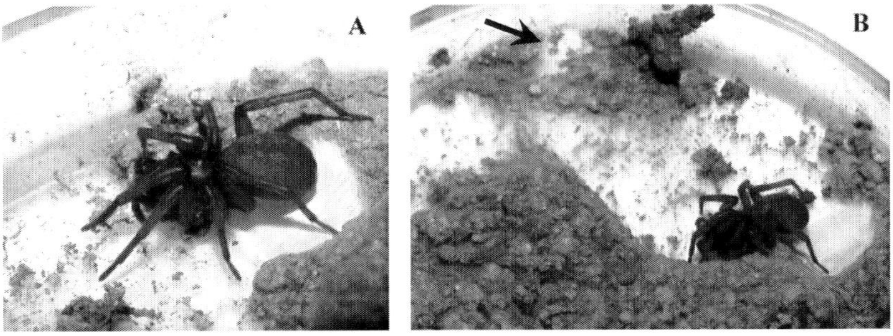

Figure 9. Female of *Haplodrassus severus* (Gnaphosidae) laying its eggs on silken basal surface of the cocoon (A) and covering the eggs with the upper layer (B). The arrow in B indicates another cocoon covered with substrate fragments. Photos by J. Benhadi-Marín.

5. Mobility and Dispersal of Spiders

When spiderlings emerge from the cocoon and are ready to search for food (Figure 10), the newly independent juveniles commonly use a mechanism known as *ballooning* to disperse. The spider searches for a high place and puts the abdomen pointing upwards; then it produces silk until an airstream lifts it and carries it to another place. This process is very useful and is also used by some adults, although there is a high probability of landing on inappropriate places. On the other hand, this process allows individuals to go over physical obstacles that otherwise would not be possible. Often, spiders also build silk bridges between close obstacles to move (Bristowe, 1958; Turnbull, 1973).

Silk is a resource commonly used by web building spider species or groups that are not good runners (due to their morphology), but wandering spiders have locomotor appendages adapted for running long distances and climb easily (Turnbull, 1973).

A peculiar aspect of the movement is the production of a security thread in response to danger stimulus mediated by a danger. Even from great heights, an individual can jump off quickly by fixing one end of the silk fiber to the substrate and sliding down in order to escape. Moreover the spider does not cut this security thread and can climb it back to its original position (Morse, 2007).

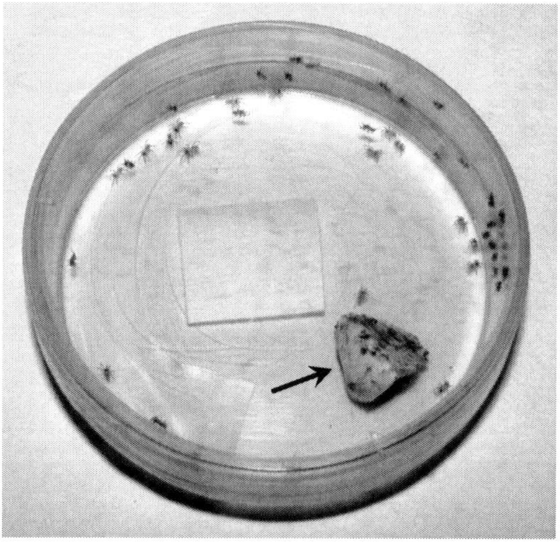

Figure 10. Several spiderlings emerging from a cocoon (indicated by the arrow) of *Haplodrassus severus* (Gnaphosidae). Photo by J. Benhadi-Marín.

6. Mortality of Spiders

The number of eggs laid during the life of a female far exceeds the need to replace parents, so it is easy to assume that there is a high rate of mortality in juvenile stages. The magnitude of mortality is determined by factors such as predation, cannibalism or lack of a suitable place that meets the nutritional needs of spiderlings once the nest or the mother is abandoned. Even in adulthood, spiders have many natural enemies such as parasites, specialist parasitoids and predators including birds, rodents and reptiles, but simply facing their winter can increase the mortality (Bristowe, 1958; Turnbull, 1973).

7. Spiders as Predators

7.1. Community Structure

Spiders species assemblages have been seen of limited interest when considering community structure because they represent only a node within a complex system (Wise, 1993). Moreover, biodiversity is an ambiguous concept without an universal measure or unit and cannot be considered a single attribute to assess it (Moreno, 2001). Hence the first and more intuitive aspect to be exploited is probably the distribution of the abundance of species along time and space. This can help to uncover possible trophic relationships with other species of the community. But even this way may represent a hard way to go due to the high diversity of spiders. Instead, the approach of using groups of species with similar characteristics (guilds) to understand how spiders play their roles as predators is often more manageable and practical, thus helping to generalize about the role of spiders in the real whole community (Wise, 1993). The functional approach and the guild concept will be focused later in this chapter.

In general, the numerical relationships between the abundance of species in a community seems to follow a pattern where usually some species are much more abundant than the rest, a greater number of species present intermediate abundances and finally, a higher number of species are usually represented by few individuals. These three groups could be called dominant, influential and accessory species respectively (Luczak, 1960, 1963). A similar classification can be extended to spiders inhabiting the agroecosystem, calling 'agrobiont' spiders to those that dominate the community and 'agrophile' to the other ones (Samu & Szinetár, 2002). A dominant species represent 50% of abundance in the whole community, as influential are those with 15-25% of abundance, with the others being accessory. However, each community must be studied and quantified individually in order to establish the threshold between categories (Luczak, 1963). These relationships between the abundance of groups were described and related thorough mathematical models in the Fisher logarithmic series and appear in a large number of communities (Fisher et al., 1943).

In terms of body size, the log-normal function usually represents a community considering the number of species and the body size (see Růžička, 1990). Body size is an useful parameter for investigating niche use and fitness due to its correlation with body length, a measure relatively easy to obtain. Size can be measured as wet or dry weight and Sage (1982) proposed a simple regression model to calculate the weight of spiders from a body size ranging between 3.0 and 35.4 mm:

$$\ln Y = a + bX + b^1X^1$$

were Y is the weight (g), X is the body length (mm) and a, b and b^1 are the regression coefficients, -6.504, 0.45919, -6.8653×10^{-3} and -7.849, 0.49335 and 8.0448×10^{-3} for wet and dry weight respectively. Body weight can either reflect changes in the environment (e.g., lack of food or water) or cause a different response to environmental changes or pressures. For example, Benhadi-Marín et al. (2016), found that body size (measured as dry weight) influenced significantly the survival probability of *Araniella cucurbitina* (Clerck, 1757) (Araneidae) when exposed to kaolin.

Růžička (1990) found that three groups of spiders could be considered in terms of body size when regarding the epigeic spider community in Czechoslovakia. The three groups were formed by spiders with about 1 mm, 1.5 mm and 3 mm of cephalothorax length respectively (Figure 11). The first group was represented mostly by Linyphiidae and the third one by Lycosidae. Within the agroecosystem context, Birkhofer et al. (2013) proposed three different clusters of body size based on the most common species found in central European cereal fields: < 4 mm (small), 2-9 mm (small to medium) and > 9 mm (large), and three hunting categories based on hunting behavior, namely: (1) ground and vegetation layer using sheet webs, (2) vegetation layer using aerial web and (3) ground and vegetation layer hunting freely.

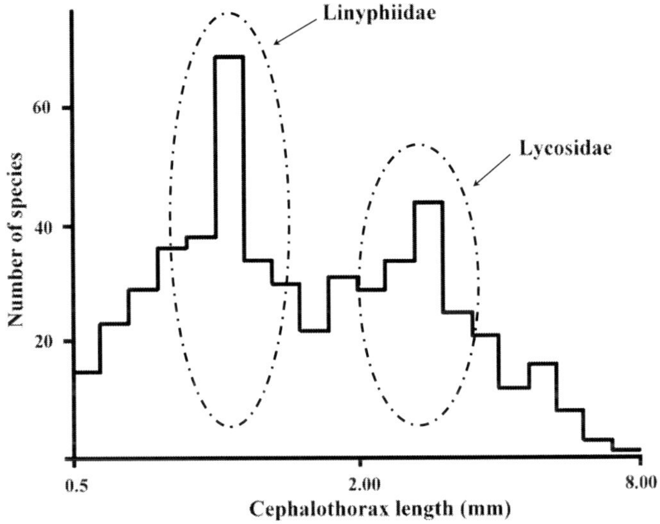

Figure 11. Relationship between species richness and body size (cephalothorax length) for an epigeic spider community. Simplified from Růžička (1990).

The different groups of spiders seem to appear distributed along horizontal, vertical and temporal gradients (Luczak, 1966) as a strategy to avoid competition (Tretzel, 1955). These gradients are reflected on several hunting strategies that are different among families (Wise, 1993):

- The Araneidae family builds orbicular webs with tiny sticky droplets and usually waits for prey in the web center;
- The Tetragnathidae family also makes orbicular webs but close to water bodies;
- The Uloboridae family weaves orb webs using special cribellate silk.
- The Theridiidae family makes irregular tangle webs;
- The Linyphiidae family uses sheet webs without any sticky silk;
- The Agelenidae family uses a funnel-like sheet web waiting in the deep end for prey;
- The Thomisidae family commonly ambushes their prey on flowers, leaves and trunks;
- The Lycosidae family usually roams endlessly pursuing and ambushing a prey;
- The Saticidae family stalks their prey using its high visual capacity;

- The Oxyopidae family runs on stems and leaves of plants searching for prey;
- The Clubionidae family hunts during the night commonly on leaves.

Overall, such spectrum of hunting strategies results in a reduction of competition for resources allowing the coexistence of related spider groups (Turnbull, 1973; Uetz, 1977). The distribution of the different hunting strategies along time and space make the assemblage of spiders an interesting group in pest management due to the possibility of several predators can prey the same pest.

7.2. Hunting Strategies and Feeding Habits

All spiders produce silk, but not all spiders use it to build webs. This feature is of major relevance and the baseline when grouping spider families regarding different hunting strategies and feeding habits. In general, two groups of spiders can be separated according to their hunting strategy: (1) those spiders with a relatively sedentary life that use silk as a hunting tool and (2) those that actively seek their prey on the substrate (Turnbull, 1973; Wise, 1993). Regarding the first group, it is important to note that a sedentary lifestyle does not exclude a high plasticity when the substrate is chosen to build the web, the prey selection and even the morphology of the web itself. In addition, the web type is relatively constant at family level, but the web architecture and the final shape can vary between genera and can even vary slightly between individuals of the same species (Turnbull, 1964, 1973).

The hunting place is influenced by multiple factors such as temperature, humidity, airflows, hours of sunshine and presence or absence of an appropriate structure to build the web. Consequently, the spider can consider a particular place as appropriate and, after a period without capturing prey, it can abandon the site (Cherrett, 1964; Jones, 1941; Tretzel, 1955; Turnbull, 1973).

The potential prey that can be captured by a spider is often limited by the characteristics of the prey such as body size and shape, lifestyle, ability to detect the web, presence of intimidating colors and production of unpalatable substances that allows it to avoid webs, escape them, or be discarded by the spider (Nentwig & Wissel, 1986; Turnbull, 1973).

The web acts as an extension of the sensory system of the spider and can have three functions: (1) an alarm system against hazards, (2) an enhanced system for catching prey that can extend beyond the perception range of the spider and (3) a system that allows to place the spider in a position of advantage over the prey, thus increasing the hunting efficiency (Turnbull, 1973). In summary, the effectiveness of web-builder spiders depends mostly on the spider species, the web design and the prey characteristics.

Species from the second group of spiders, the active hunters, shows variable levels in their itinerant habits and several factors come into play when using their hunting strategy. Depending on the development of the eyesight, length and strength of the legs, tarsal sensory organs and ability to mimicry, different ways to find and chase a prey can be observed (Turnbull, 1973). This group of spiders exploits similar food resources but using different hunting techniques. Some species ambush their prey using its ability to mimicry (e.g., *Thomisus onustus* Walckenaer, 1805 (Thomisidae)), while others take advantage of their high visual power to chase and attack their prey directly (e.g., Lycosidae, Pisauridae and Salticidae). Several species take advantage of their secondary eyes (with reverse view) and are nocturnal hunters (e.g., Anyphaenidae, Clubionidae and Gnaphosidae) (Berland, 1938; Foelix, 1996; Turnbull, 1973).

Regarding the feeding habits, spiders are generalist predators and produce poison (with some exceptions, such as the Uloboridae family), which is inoculated into their prey through the chelicerae fangs as part of hunting. Notwithstanding, this is not the only function of the chelicerae as they are also used as a tool to transport objects, hold and manipulate prey and build nests (Berland, 1938). However, the spider esophagus is not able to accept solid food, therefore spiders cover the body of its prey with digestive enzymes and it will not be ingested until the tissue has been digested. Moreover, some spiders use the maxilla to grind the food. Once the tissues are digested, the resulting mass is ingested by suction through a pharyngeal pump (Turnbull, 1973).

7.3. Spiders as Natural Enemies of Pests

The suppression of pests by predators has been a subject of interest for agricultural managers within sustainable farming practices. Traditionally, the equilibrium between predators and prey populations is essential to achieve good results for biological control (Murdoch et al., 1985) especially from a reductionist point of view where a single predator or a single prey species are considered (Riechert, 1999). Also, damage thresholds are frequently exceeded despite spider predation (Bishop & Blood, 1981). This suggests that spider communities are unable to control pests, however, agroecosystems are first of all 'systems' (often complex systems) and great efforts have been made in order to model the role of spiders predation in agroecosystems (e.g., Chatterjee et al., 2009). Mathematics rarely captures the whole dynamics between all components of the network thus doing unreliable or even unpredictable the result of the interactions between its elements. In addition, biased measures of species community composition and predation rates may lead to underestimate the potential of spiders as natural enemies.

Nonetheless, successful examples involving relatively simple systems (e.g., habitat manipulation and laboratory experiments) of biological control provided by spiders are mentioned in literature. For example, considering the economic relevance of wheat, several authors studied the effect of spiders against its pests, mostly focusing on aphids. Chiverton (1986) found that linyphiid spiders together with other predators such as carabids and staphilynids (Coleoptera) reduced the population of *Rhopalosiphum padi* (Linnaeus, 1758) (Hemiptera: Aphididae) in spring barley in Sweden. Kuusk et al. (2008) found that *Pardosa* C. L. Koch, 1847 spiders (Lycosidae) also fed on *R. padi* and Oelbermann & Scheu (2009) also demonstrated the benefits of the presence of *Pardosa* in wheat, both in terms of aphid predation and plant performance. More recently, Chapman et al. (2013) provided a molecular framework to study the feeding habits of an epigeal spider community in a winter wheat agroecosystem.

Regarding other crops, Ghavami (2008) suggested that the salticid species *Thyene imperialis* (Rossi, 1846) could be a potential natural enemy of key cotton pests in Iran in laboratory assays. Also, Shivakumar & Kumar (2010) indicated lynx spiders (Salticidae) as potential natural enemies of *Spodoptera litura* (Fabricius, 1775) (Lepidoptera: Noctuidae) and Wyss et al. (1995) observed that web-building spiders, such as *Araniella* (Araneidae), contributed to aphid control in apple orchards in Switzerland. Laszló et al. (2015) estimated that 24% of spider prey were pests in apple orchards and Monzó et al. (2010) used DNA markers of *Ceratitis capitata* (Wiedemann, 1824) (Diptera: Tephritidae) to show how *Pardosa cribata* Simon, 1876 (Lycosidae) played an important role in biological control predating this harmful fruit pest in citrus orchards in Spain.

Spiders have been shown to play an important role helping to reduce pest abundance peaks, maintaining their populations at constant levels rather than reacting to pest outbreaks (Riechert & Lockley, 1984). In situations of prey population peaks, spiders show a more selective behavior choosing their food and this, added to the fact that both spiders and prey occupy different strata (vertical, horizontal and temporal) in the agricultural ecosystem, may lead to the existence of a certain degree of specialization in response to these microhabitats (Maloney et al., 2003).

Prey selection behavior is not an expected characteristic in an effective natural enemy; however, pest density seems to be an important factor driving the potential killing capabilities of spiders. In fact, spiders usually kill more prey than necessary when sufficient food is available. This behavior is known as wasteful killing or superfluous killing (Sunderland, 1999) and implies a clear advantage of the use of spiders in agroecosystems as natural enemies. Samu & Biró (1993) described this effect using cursorial spiders (Lycosidae) as model organism and observed both partial consumption and non-feeding on killed flies at high prey densities. Also, Riechert & Maupin (1998) proved that different web-builders also exhibited this behavior suggesting greater feeding efficiency in partial feeding. Therefore, different functional groups, probably occupying different strata within the agroecosystem, are able to kill significant amounts of insects especially at high densities. Since the rate of food acquisition directly affects growth rate, molting and size at maturity (Uetz, 1992), it is supposed that spiders will prefer a prey-rich habitat and will move to it if they can expect to capture more prey than required to survive and reproduce (Caraco & Gillespie, 1986).

Although the physiological mechanism underlying superfluous killing in spiders has not been explained yet, the consequences for prey population are clearly relevant. Several examples on how different pest populations respond to spider predation can be found on comprehensive reviews done by Greenstone (1999), Riechert (1999), Sunderland (1999) and Maloney et al. (2003). Despite the high amount of available reports on the role of spiders acting as natural enemies, pest suppression strategies based on spiders have been rarely implemented. One possible reason is the development of a complex integrated protection strategy that involves social, cultural and political interests (see Ehler, 1998) using a tool that implies high complexity itself, i.e., the assemblage of spiders.

7.4. The Functional Approach: What, Why and How

Beyond the role and identity of a single or a group of species in an ecosystem and its/their responses to environmental changes, groups of similar species and characteristics of individuals are currently in the spotlight within the so called functional ecology. This approach to, species roles and dynamics in nature and their link to ecosystem services provision is based on 'functional traits.'

Functional traits can be seen as key characteristics by which single species and groups of species influence ecosystem properties (de Bello et al., 2010) and are defined as a feature of an organism, which has demonstrable links to the organism's function (Lavorel et al., 1997). Consequently, 'functional groups' of species (usually known as 'guilds,' see Root (1973)) can be defined as an assemblage of organisms with similar functional trait attributes (Harrington et al., 2010), or in other words, species using the same class of resources in a behaviorally similar way (Simberloff & Dayan, 1991).

Among animals, spiders have been suggested as good candidates for ecological studies thorough a functional approach and appropriate for all kinds of biomes and habitats (Cardoso et al., 2011). A first advantage of this framework is related to taxonomy itself. Both experienced and novice researchers have faced the hard way to deal with spider species identification. Currently, there is a lot of taxonomic literature as well as great on-line tools and comprehensive databases concerning spider's diversity and taxonomy. However, accuracy on species-level identification involves careful and time consuming genitalic analysis and remains impossible in the case of immature specimens. The common way to solve this situation is to turn to an expert in spider taxonomy but obviously this is not always possible (see Young & Edwards, 1990).

Nevertheless, using higher taxonomic ranks, such as family, as the baseline to construct functional groups (assuming that species perform in a similar way) can greatly facilitate the task. On the other hand, different species may play the same role in the ecosystem, overlapping in phenology, size and prey taken e.g., two big orb-web spiders *Argiope aurantia* Lucas 1833 and *Argiope trifasciata* (Forskål, 1775) (Araneidae) coexist and exploit the same resources (Enders, 1974) and, in such case, the effect of spiders on prey population should be considered as a whole, and here is why a functional approach becomes especially useful for biological control.

Defining reliable functional groups is, however, hard due to the high diversity of these arthropods and some problems need to be addressed. Firstly, the behavior of an extremely high percentage of spiders remains currently unknown and it is commonly assumed as being homogeneous among the member of the same family or genus (which may not be real). Secondly, many spider guilds are usually based on the foraging strategy. But, different strategies may target similar prey and similar strategies may be directed towards different prey (Cardoso et al., 2011); in addition, as generalist predators, knowledge about spider's prey range and preferences is scarce. Thirdly, functional traits commonly used such as body size, color and phenology can be extremely variable even within populations and follow interannual variation, thus becoming uninformative.

Hence, Cardoso et al. (2011) suggested four criteria that need to be considered in order to approach the challenging goal of defining proper functional groups according to the hypotheses to be tested in a study: (1) the objectives of the study, (2) the spatial scale of the study, (3) the taxonomic scale and (4) data available. Summarizing, the defined guilds should reflect the natural history of its members but a certain degree of laxity has also to be assumed (Uetz et al., 1999).

The guild concept in spiders has been widely investigated, developed and used to test the effect of environmental changes on spider assemblages. Balogh & Loksa (1948) considered three groups of spiders namely, web spinners, cursorial and saltatorial spiders; however the most basic and largely used division corresponds to the two extremes of hunting strategies identified by Pianka (1966) and Schoner (1971), i.e., "sit-and-wait" and "mobile" predators. Uetz (1977) defined these two groups as 'web-builders' and 'wanderers respectively' and Post & Riechert (1977) split them into 11 groups according to the web structure and circadian activity.

Since the study of predation on insects is commonly directed toward pest suppression, several works addressed the issue of the role and composition of the guild of spiders in agroecosystems. Nyffeler (1982) considered three guilds considering the role of spiders as insect predators in abandoned grasslands, meadows, and cereal fields. Riechert & Lockley (1984) defended the potential of spiders as natural enemies on a review dealing with eight guilds mentioned along literature, and Young & Edwards (1990) distributed the spiders of North America between five guilds extracted from Gertsch (1979) and Comstock (1940). This lack of homogeneity between studies was somewhat solved by Uetz et al. (1999) in a classic work where information concerning spiders commonly found in agroecosystems was hierarchically clustered and eight new functional groups (foliage runners, running spiders, ground runners, stalkers, ambushers, sheet web-builders, wandering sheet/tangle-weavers, orb-weavers and space web-builders) were described based on earlier assignments providing a framework for subsequent studies. Unfortunately, the problem of the diversity of spiders all around the world still remained and, more recently, Cardoso et al. (2011) undertook the task at a global scale of redefining eight new guilds (sensing, sheet, space and orb web weavers and specialist, ambush, ground and other hunters) that encompassed the whole diversity of spiders at the family level. Thus, it was developed a powerful tool allowing an easy application of the guild concept thorough few functional groups that turns available the comparison between studies of different regions.

The functional approach is especially useful in agricultural research when morphological, physiological, and behavioral traits are regarded as surrogates for ecological functions. Moreover, regarding spiders as generalist predators it becomes an important tool helping to test hypotheses involving niche overlap, intra-guild competition, prey selection, functional responses, etc., and their consequences on the role and effectiveness of spiders as natural enemies.

8. Effects of the Most Common Agricultural Practices on the Biodiversity and Life-History Parameters of Spiders

Spiders are one of the most abundant predators in agroecosystems. For example, in Europe, the overall mean density of spiders have been estimated in ≈ 80 spiders/m^2 (Nyffeler & Sunderland, 2003). Basic functional groups (i.e., web-builders and active hunters) are represented in agroecosystems and in all vertical strata (Nyffeler & Sunderland, 2003). Agroecosystems seem to sustain characteristic spider communities and dynamics. Agrobiont spiders should be indicators of arable habitats and may be able to synchronize their lifecycle with the crop-growing season. In northern temperate Europe, these key species are mostly represented by Linyphiidae whereas in USA the families Tetragnathidae, Araneidae, Linyphiidae, Theridiidae, and Dictynidae are the most representative web-builders and Oxyopidae, Salticidae, Clubionidae, Thomisidae and Lycosidae dominate within the active hunter guild (Nyffeler & Sunderland, 2003).

Land use intensification can greatly reduce species richness and ecosystem functioning (Flynn et al., 2009). Management practices such as tillage, vegetation cutting, grazing and application of herbicides, insecticides and fertilizers may alter the community of spiders and thus, decrease its effectiveness as natural enemies. On the other hand, biotic (e.g., vegetation and landscape complexity, plant richness, coverage and density) and physic/structural characteristics (e.g., water bodies, vegetation strips, stone density and stonewalls) can benefit spider assemblages by providing alternative food and shelter and may help to reduce intra-guild competition.

Hereafter we analyze the effects of the most common agricultural practices on the population of spiders in the scope of biological control methods.

8.1. Soil Management

The assemblage of spiders is vertically stratified in arable lands, in particular, in perennial crops. Hence an important part of the community inhabits the soil and is confined to it. Such as other ground dwelling natural enemies, spiders rarely climb up the tree pursuing pests (see Sunderland et al., 1997). Accordingly, any agricultural practice involving soil mechanical management can affect directly the diversity of edaphic spiders by removing microhabitats provided, for example, by stones that are used by ground dwelling spiders (e.g., Gnaphosidae) as shelter or zone for hunting or reproduction (Figure 12).

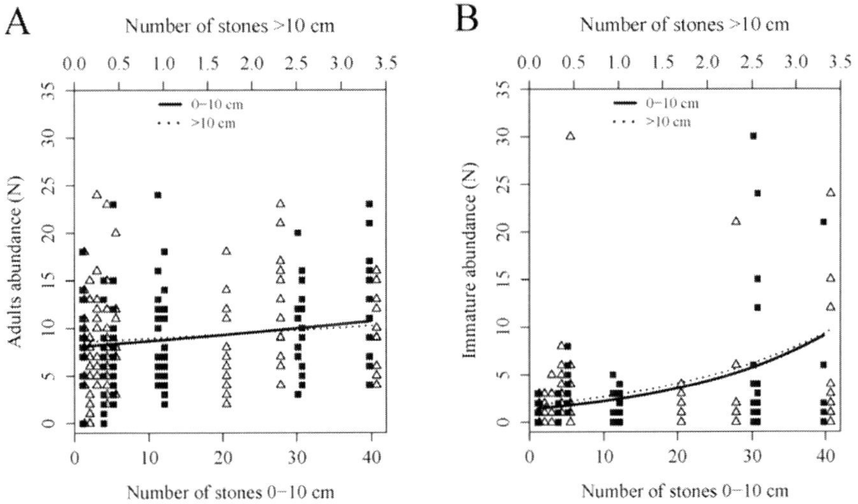

Figure 12. Influence of the number and size of stones on a ground community of spiders in olive groves in north-eastern Portugal. The total number of adults is not affected (A), whereas the number of juveniles significantly increase with the number of stones. Squares correspond to smaller stones (0 - 10 cm) and triangles to bigger stones (>10 cm). The size of stones was measured along their longer axis and the number of stones counted in 1 m^2. Benhadi-Marín et al. (data not published).

Practices such as tilling, weed control with herbicides or cutting can result in the reduction of the population of spiders (Thorbek & Bilde, 2004) and prey availability (Butt & Sherawat, 2012, see also Dicks et al., 2013). For example, Blumberg & Crossley (1983) registered similar percentages of individuals and species represented by spiders in non-tilled and old-fields whereas a substantially lower proportion was found in tilled sorghum fields; Holland & Reynolds (2003) found that the populations of Linyphiidae and Lycosidae were significantly reduced in ploughed strips in a mixed farm. Moreover, Benhadi-Marín et al. (2013) found that the reduction or absence of a suitable habitat for spiders under the trees in tilled chestnut crops might have forced spiders to migrate into the canopy and suggested that when soil cover is removed by tillage or frequently disturbed by grazing, spiders may climb the trunks and search for a new habitat or food.

Less disturbed systems favor a rich soil biota that can improve nutrient cycling and plant health (Zehnder et al., 2007) and often provide a diversification of interspersed micro-habitats (Samu et al., 1999). This, a reduction of mechanical disturbances will increase the diversity of spiders in agroecosystems. However, conservation tillage systems may actually suffer from recurring pests, lower yields, and weed problems, but the use of novel habitat manipulation techniques such as the allocation of modular artificial shelters for natural enemies can be a successful strategy in order to mitigate the effects of tillage (Halaj et al., 2000).

Herbicides are one of the most applied pesticides in agroecosystems in order to control weeds. Hence, several studies have been carried out trying to assess the effect of these formulations on the community of natural enemiespesr control agents (e.g., Geiger et al., 2010). Regarding spiders, the response of different functional groups such as ground runners and web-builders to herbicides have been investigated (e.g., Michalková & Pekár, 2009; Benamú et al., 2010; Evans et al., 2010; Pékar, 2012; Wrinn et al., 2012; Leccia et al., 2016). In general, both direct and indirect effects can be identified. Indirect effects involve changes in habitat structure. Spiders prefer rich and complex-vegetated ecosystems (Rypstra et al., 1999), hence changes in vegetation diversity due to herbicide application may affect spiders in two ways: (1) reduction of the number of webs, shelters and overheating protection sites and (2) reduction of prey availability by decreasing the amount of food for herbivores and consequently limiting prey abundance for spiders (Pékar, 2012). Direct effects involve both lethal and sub-lethal effects. Lethal effects are considered when mortality rate directly increases after exposure and sublethal effects are defined as physiological and/or behavioral changes in an individual that survived after exposure (Pékar, 2012). Behavioral changes may affect mobility, orientation, feeding behavior, oviposition, and learning (Desneux et al., 2007). Sublethal effects have been commonly found in laboratory tests, although it has been argued that herbicides do not have significant lethal effects on spiders (Pékar, 2002). Within a biological control framework, Korenko et al. (2016) found that *Pardosa*, a spider genus commonly found in agroecosystems, killed significantly less flies when exposed to different herbicides. Although spiders were not directly killed by herbicides, if a trait related to hunting success is affected, the global success of spiders as natural enemies is reduced as well. Thus, research on how correlated traits respond to herbicides is mandatory in order to effectively assess how pest control provided by spiders may shift from successful to unprofitable.

8.2. Insecticide Application

In general, spiders are more sensitive to some pesticides than many pests, such as synthetic pyrethroids, organophosphates and carbamates (El-Wakeil et al., 2013). A large number of studies demonstrated that, in many cases, the application of insecticides decreased both the abundance and richness of spiders (El-Wakeil et al., 2013; Pékar, 2012), therefore reducing the possibility of biological pest control provided by these arthropods. Fortunately, pest management based on the application of broad-spectrum synthetic insecticides has been progressively replaced by more environmentally-friendly alternatives: (1) the integrated pest management (IPM), based on the use of synthetic insecticides with acceptable environmental impact used when required and usually only at specific times, and the use of biological insecticides, and (2) organic farming, which restricts the use of synthetic insecticides and promotes the use of predators, parasitoids and entomopathogens. This change of paradigm clearly had a positive effect on the community of spiders and it has been demonstrated in tea gardens, vegetables, wheat, olive groves and fruit orchards where the diversity of spiders increased when compared to conventional management (see Pékar, 2012).

It is beyond of the scope of this chapter to provide a complete list of insecticides and its effects on different spider species. Instead, a conceptual overview on how pesticides affect spiders based on the comprehensive ecotoxicological review performed by Pékar (2012) is provided (Annex 1).

8.3. Mineral Particle Films

Numerous negative effects resulting from the use of synthetic pesticides including risks for non-target organisms and persistence in the environment have led to the development of mineral particle films as an alternative in pest control programs (Glenn & Puterka, 2005; Glenn et al., 1999). Kaolin-based products, one of the most used mineral particle films, can act as repellent or barrier for pests, affecting the recognition and attractiveness of host plants (Showler, 2002) and it has been already used in several crops such as pear, apple, olive-, walnut and almond, citrus, wine grape, pecan, cotton, tomatoes and cabbage (see Benhadi-Marín et al., 2016). However, knowledge about its possible effects on beneficial arthropods and especially on spiders is still scarce. Sackett et al. (2007) found that kaolin application in apple orchards reduced the relative abundances of jumping (Salticidae) and crab spiders (Philodromidae), whereas the abundances of web-building spiders such as Araneidae, Dictynidae and Theridiidae were not affected. Also in apple orchards Markó et al. (2010) observed that *Xysticus* L. Koch, 1835 (Thomisidae) and *Theridion mystaceum* L. Koch, 1870 (Theridiidae) populations strongly decreased when kaolin was applied. On the contrary, *Dictyna uncinata* Thorell, 1856 (Dyctinidae) and *Clubiona* Latreille, 1804 (Clubionidae) tolerated kaolin applications. A plausible explanation about the different effects of kaolin application on the population of spiders may be related with the lifestyles of different functional groups, e.g., if kaolin is applied during the day, nocturnal spiders such as *Clubiona* can avoid direct exposure to the mineral particle (Markó et al., 2010). Notwithstanding, spiders can be exposed to kaolin directly during its application by forming a particle film on the body and/or by ingesting particles when feeding on sprayed prey or substrate surfaces. Benhadi-Marín et al. (2016) used the web-builder *A. cucurbitina* as model species in order to assess how kaolin affects spiders in laboratory tests and found that kaolin significantly reduced the survival of this species when applications were done (1) on the substrate surface and (2) on spider and prey body together, registering a survival reduction of 48% and 56%, respectively. As in the case of synthetic pesticides, functional trait-related characteristics such as web-building ability, legs adapted to running and body size seems to be closely related to the final effect of kaolin application on the life span of spiders.

9. Influence of Landscape on Populations of Spiders

Spiders are not distributed randomly in ecosystems and spatial and temporal patterns have been repeatedly observed and reported in the literature. Agroecosystems are not the exception and probably one of the most obvious patterns of the diversity of spiders is the dominance of few agrobiont species that are adapted to cyclical disturbances caused by agricultural practices. The probability of survival to regular disturbances relies on migration and recolonization skills and spiders proved to be successful travelers and colonizers by moving on the ground or by air thorough ballooning (Marc et al., 1999). Spiders also seemed to be more abundant in perennial crops than in other agroecosystems during winter (Schmidt & Tscharntke, 2005); accordingly, the existence of overwintering sites can be one determinant factor for biocontrol since it is desirable to maintain the population of spiders in crop fields during pest outbreaks and, therefore, immigration from surrounding habitats can be necessary (Pywell et al., 2005). Reproduction, survival and dispersion dynamics involve landscape, habitat and micro-habitat levels (Samu et al., 1999) and spider fitness can be affected in different ways by the available resources provided by each level of the spatial hierarchy.

9.1. Landscape Scale

The study of the effects of landscape on wildlife has been commonly studied from two different perspectives because of the difficulty of working in large areas (Samu et al., 1999). The first one consists on modeling the species distribution on the landscape based on biological knowledge using different scenarios. Topping (1999) developed individual agent based models and found that the abundance of spiders did not depend on the landscape spatial distribution; nevertheless, when habitat patch size increased, maintaining a constant habitat area, the size of the population of spiders increased as well. Also, the simulations suggested that high juvenile dispersal was the optimal dispersal strategy for spiders.

The second approach is based on field experiments at a small landscape scale, usually considering patch fragments such as adjacent crops and/or wild areas and its margins. Öber et al. (2008) found that the abundance of lycosids increased during pest establishment, thus these ground runners were recolonizing the crop from adjacent patches; however, abundance did not depend on the landscape heterogeneity whereas the latter was positively associated with lycosid species richness. On the other hand, dispersal based on ballooning resulted in a quick recolonization by linyphiids and this family did not show an increase in activity, density or species richness over time. Landscape may influence not only dispersal activity or behavior but also spider fitness. Large fields of annual crops originated high body condition in *Pardosa* females due to less competition for resources within homogeneous areas, whereas farming practices had no effect on either body condition or fecundity parameters (Öberg, 2009).

Landscape can also affect spiders at smaller scales. Schmid et al. (2007) suggested that a 95 m to 3 km radius around the study fields potentially reflects dispersal distances (depending on the species) and in general, local species richness enhanced due to the presence of non-crop habitats. Even at a smaller scale, Sackett et al. (2009) found that colonization rate and foliage-dwelling spider assemblages were affected by distances ranging from 10 to 50 m in apple orchards.

9.2. Habitat Scale

Agricultural fields can typically be viewed as habitats composed by homogenous vegetation with a number of micro-habitats within a delimited area. Hence, it is supposed that farming practices alter the diversity of micro-habitats within crops, for example, topography and soil cover (e.g., plants, organic matter and mulching) may be important factors driving the community of spiders (Samu et al., 1996; Samu et al., 1999). In fact, Altieri & Schmidt (1986) caught significantly more spiders in a cover-cropped orchard than in a tilled orchard using pitfalls and addition of mulches or manures to the soil surface provides a more complex and diverse habitat giving spiders protection from natural enemies and improving microclimate conditions (Sunderland & Samu, 2000).

The number of species in weed strips and field margins is usually higher than in the adjacent fields. Richly dense structured vegetation can serve as overwintering place and, accordingly, Nentwig et al. (1998) found significantly higher spider richness in older weed strips than in younger strips. In a similar way, the more diverse plant structure found in field edges can accommodate a more complex spider community regardless the agroecosystem type and management due to a reduced management intensity and higher immigration (Batáry et al., 2003). For example, field margins provide an excellent source habitat for *T. tenuis* acting also as shelter for many other beneficial invertebrates (Bell et al., 2002). On the contrary, Pfister et al. (2015) found that the abundance of the orb web builder *Mangora acalypha* (Walckenaer, 1802) (Araneidae) decreased from field centers towards the hedgerows probably due to interspecific competition avoidance and a higher pest population which may serve as prey. In general, the increasing spider densities towards field margins can be explained by the preference of spiders for seminatural habitats that provide overwintering sites, food, better microclimate conditions, and more complex vegetation structure (Pfister et al., 2015).

The presence of water bodies can also enhance the diversity of spider, especially considering species linked to watercourses or ponds such as *Tetragnatha* Latreille, 1804 (Tetragnathidae) that can walk on the water surface (Foelix, 1996). Pfister et al. (2015) found that the abundance of *Tetragnatha* cf. *montana* Simon 1874 increased towards both hedgerows and riparian margins whereas *Tetragnatha extensa* (Linnaeus 1758) increased only towards riparian margins. Interestingly, flooding episodes does not seem to affect the community of spiders (Lafage & Pétillon, 2016) probably due to their ability to survive to periods of submersion (e.g., Pétillon et al., 2016) or their capability to deal with encounters on water (Hayashi et al., 2015).

9.3. Micro-Habitat Scale

Spider abundance is determined by site selection, by rates of survival and reproduction related with the site and site abandonment, being in turn determined by abiotic and biotic factors (Samu et al., 1999). There is strong evidence that physical spatial structures given by vegetation complexity can be the most important abiotic factor for spiders determining the habitat quality at small scale especially for web builder functional groups. Web builders need not only appropriate attachment points but also the appropriate space arrangement to construct their webs and those are provided by a complex spatial distribution of plants. In addition, prey richness also increases with plant diversity and vegetation cover being lower at tilled sites when compared with non-tilled sites (Diehl et al., 2013). Grass and sub-shrub covers seem to be determinant for spider species richness. However, it is necessary to take into account that relationships between the diversity of spiders and vegetation could not be direct and may depend on other abiotic factors (Schaffers et al.,

2008). For example, temperature influences vegetation structure (Jiménez & Lobo, 2007).

Biotic factors include vibrational, olfactory and visual cues related to prey, and spiders have shorter residence times in micro-habitats where prey is scarce compared with sites where food is abundant (Samu et al., 1999). In this situation, the occurrence of alternative food is relevant if it provides nutrients that can increase the lifespan of spiders. When insect preys are scarce, aerial plankton is of vital importance especially for spiderlings, and the first instars may increase the proportion of non-prey food consumed using insects as supplementary resource (Smith & Mommsen, 1984). Non-prey food includes pollen, nectar and honeydew. Accordingly, Eggs & Sanders (2013) reported that about 25% of juvenile Araneidae orb-weaving spiders diet consisted of pollen. Microorganisms growing on the proteinaceous silk and pollen deposited on the sticky drops may be sufficient to nourish spider of the first instar (Smith & Mommsen, 1984). Moreover, adults of the orb weaver *Gasteracantha cancriformis* Linnaeus, 1758 (Araneidae) have also been observed eating pollen from the corbicula of the leg of a bee (Gregory, 1989). A different web builder guild such as Linyphiidae (sheet web builders) also consumed pollen consistently during laboratory assays (Peterson et al., 2010) revealing that ground living web builder spiders can also benefit from this resource.

Spider webs were very efficient in trapping pollen particles (Bera et al., 2002), but web builders are not the only functional group that benefit from pollen. Vogelei & Greissl (1989) fed spiderlings of the ambusher *T. onustus* in laboratory and those feed on pollen increased significantly their mean lifespan compared with starved spiderlings. Suetsugu et al. (2015) reported *Clubiona*, a foliage runner spider, removing pollen from flowers of the nectariferous orchid *Neottianthe cucullata* (L.) Schltr. 1919.

Flowers also provide nectar, an alternative source of amino acids, lipids, vitamins and minerals. Jackson et al. (2001) confirmed that 90 species of salticids fed on flowers. Suetsugu et al. (2015) suggested that *Clubiona* also consumed nectar when visiting orchids and Vogelei & Greissl (1989) found that artificial nectar (i.e., sucrose 30%) increased significantly the longevity of *T. onustus*. Taylor & Pfannenstiel (2008) collected *Cheiracanthium inclusum* (Hentz, 1847) (Miturgidae), *Hibana futilis* (Banks, 1898) and *Hibana arunda* (Platnick, 1974) (Anyphaenidae) in cotton fields and used the cold anthrone test to detect fructose in their guts and results were positive for both adults and immatures indicating consumption of extrafloral nectar. The same test was used by Chen et al. (2010) showing that *Ebrechtella tricuspidata* (Fabricius,

1775) (Thomisidae) and eight more families (Oxyopidae, Pisauridae, Salticidae, Lycosidae, Tetragnathidae, Araneidae, Nephilidae and Agelenidae) also consumed nectar. Sexual dimorphism is commonly accentuated in crab spiders and adult males rarely feed. Pollard et al. (1995) suggested that males of *Misumenoides formosipes* (Walckenaer, 1837) drink nectar in order to replaced fluid loss through evaporation.

Benhadi-Marín et al. (data not published) tested the effect of alternative food on the ground dwelling spider *Haplodrassus severus* (C. L. Koch, 1839) (Gnaphosidae) including honey 10%, glucose (0.5 M), pollen (10%) and honeydew of the black scale, *Saissetia oleae* Olivier, 1791, a secondary pest of the olive tree. Each treatment gave significantly longer longevities than control (water). Pollen provided the lowest longevity while the spiderlings that fed on the black scale honeydew almost triplicated the mean longevity reached with the negative control (Figure 14). Spiderlings fed on sugars probably obtained energy but lacked proteins while the opposite could happen with pollen. On the contrary, honey and honeydew probably provided both sugars and essential amino acids giving significantly longer longevity to *H. severus.*

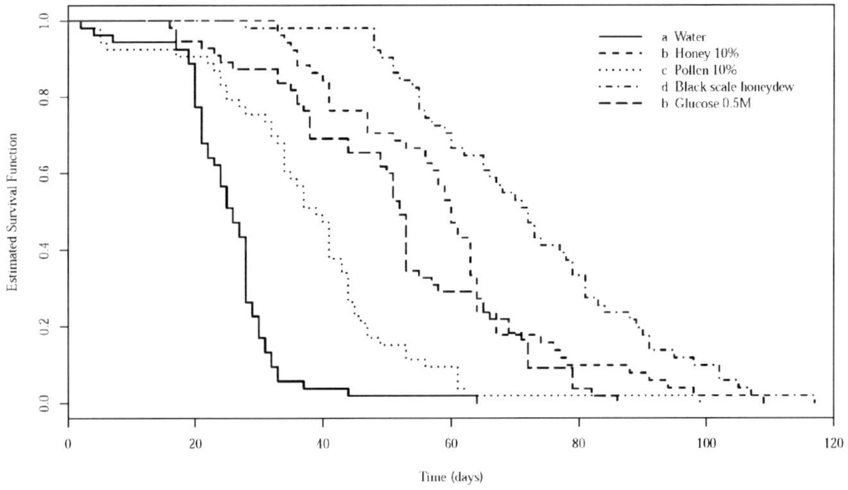

Figure 14. Survival curves of *Haplodrassus severus* (Gnaphosidae) fed on non-prey foods. Different letters in the legend indicate significant differences between treatments.

Energetic costs and risks associated to hunting are supposed to decrease in spiders feeding on flowers; moreover, nutrients, venom and digestive enzymes necessary to capture and digest a prey are also saved (Jackson et al., 2001).

Probably, each non-prey food alone is not enough to fulfill the nutritional requirements of an individual, but spontaneous plants (flowers) and honeydew may provide important nutrients to sub-adult spiders.

10. Spider Venoms as Bioinsecticides, a New Perspective

The interest in developing new and safer insecticides has been increasing during the last decades mostly due to legislative directives regulating the use of chemical insecticides. Their use still dominates in pest control strategies, but their spectrum of activity is still too wide with high levels of toxicity for non-pest species (Windley et al., 2012). Governments banned the use of first-generation insecticides characterized by their poor selectivity thus significantly reducing the available spectrum of insecticides (King & Hardy, 2013). In this context, natural products raised as the new paradigm. From this point of view, the potential of biocompounds could not be seen directly as a source of viable insecticides but as a source of molecules with biological activity which can be maximized by the synthesis of the analogues (Hodgson, 1984).

In recent years, research has been directed to mimic the natures' way to control pests. Several works have been focused on venomous insect predators or parasitoids searching for natural biocides naturally present in arthropods such as mites, scorpions, spiders, and wasps (Quicke & Usherwood, 1990, Windley et al., 2012). However, bioinsecticides should not be seen exclusively as products able to kill pests. Their development must minimize persistence of adverse metabolites in the environment and maximize their selectivity towards a target pest (Nentwig, 1993).

Spiders are mostly generalist predators and it would be reasonable to think that their venoms are suitable to subdue a wide range of prey (with the exception of stenophagic species, such as ant-eating spiders, that usually do not include agricultural pests on their diet). Moreover, spiders inject venom in their prey thorough the chelicerae fangs, an impossible technique to apply in regular pest control in the field. Finally, nothing is known about the persistence of spider derived compounds in the field. These facts together may lead to consider spider venoms as bad candidates to bioinsecticides. Nevertheless, there is a high interest in spider venoms nowadays because they comprise an extensive library of potent insecticidal neurotoxic peptides that are able to target novel sites in insects (Blagbrough et al., 1992; Windley et al., 2012).

Currently there are 1405 curated toxin records from 97 spider species available on ArachnoServer, a database containing information on the sequence, structure and biological activity of proteins derived from spider venoms (Herzig et al., 2011) and there is a long way to the validation of merchantable products. In addition, as summed up by Windley et al. (2012) prior to consider a spider-venom peptide as a safe novel pesticide several conditions need to be met, namely: (1) to present a broad pest specificity (2) to have low toxicity in non-target organisms, (3) to remain in the environment long enough to be effective, but not too long to induce development of resistance, (4) to be inexpensive to produce, (5) to be easy to formulate and deliver, (6) to be able to publicly be perceived as innocuous and (7) to be accessible to end-users (e.g., farmers).

It is actually hard to meet all those conditions, nevertheless a continuous and increasing effort is emerging in this area and currently there are also several possible commercialization ways, such as the development of recombinant baculovirus or a transgene encoding spider toxin into an entomopathogenic fungus, orally active acaricidal and/or insecticidal agents and the design of conformational constrained non-peptide mimetics to be used as foliar sprays (Windley et al., 2012).

Pesticides can affect arthropod behavior and also exert a range of sublethal effects that may not be detectable at field scale. In the case of bees, important aspects of behavior related to foraging and survival, such as learning and memory can be affected after exposure to pesticides and should be investigated as well (Palmer et al., 2013; Decourtye et al., 2004a, 2004b). A successful spider-venom based bioinsecticide was recently tested by Nakasu et al. (2014) using the fusion protein technology for linking insecticidal peptides to a plant lectin 'carrier' protein. The venom of the Australian funnel web spider *Hadronyche versuta* (Rainbow, 1914) (Hexathelidae) was used as peptide source and tested on honeybees (*Apis mellifera* Linnaeus, 1758) using three exposure routes: (1) injection (emulating a worst case scenario), (2) oral and (3) contact assays; no adverse effects were detected even when bees were directly injected with the fusion protein.

Finally, the Vestaron Corporation recently received the U.S. EPA approval to remove the bee toxicity warning statement from its spider-venom based biopesticide SPEAR™ label (Vestaron, 2016). In conclusion, it seems that the future of development and use of biopesticides based on spider venom is a promising research area and will occupy an important place in the market of insecticides in the future years.

11. Spiders in the Olive Grove Agroecosystem

The olive tree (*Olea europaea* L. 1753) is a crop with high importance in the Mediterranean Basin where 98% of the global production of olive oil is located. In southern Member States of the EU, it has a multiple importance particularly in what concerns to economic, social, cultural and landscape aspects.

In recent years, strong market pressures have encouraged growers to intensify production, resulting in the introduction of new olive planting systems (Estevão et al., 2012; Tous et al., 2011). These changes are of utmost concern because of their negative impacts (i.e., pollution and degradation of natural resources, loss of biodiversity, insecticide resistance and residues in olives) (Hawkes et al., 2005; Soares et al., 2004). Notwithstanding, the European Union (EU) has supported the adoption of environmentally-friendly farming practices through the agri-environment schemes. Moreover, the EU has established an ambitious strategy to halt the loss of biodiversity and ecosystem services through the promotion of sustainable agriculture and more use of green infrastructures, by 2020 (European Commission, 2011). As an agroecosystem with high ecological potential, olive crop is a very promising cultivation for implementing environmental friendly management strategies specially where traditional olive groves are still major sources of income and employment.

In response to an increasing concern with agricultural practices in olive groves, different tools have been developed to study and evaluate the sustainability of agroecosystems such as the use of bioindicators (Lawes et al., 2005, Longcore, 2003). Spiders have been indicated as potential bioindicators of disturbance satisfying the criteria established by Mulhauser (1990): (1) there is previous information about the organism, (2) easy field re-encounter of the group to which the organism belongs, (3) easy recognition of the organism and (4) the use as many groups of bioindicators as possible in order to obtain a global perception of the ecosystem. Accordingly, the potential of Araneae within biological control schemes and bioindication resulted in an increasing research on the role of spiders in the olive grove agroecosystem. In general, the biodiversity of spiders follows an increasing pattern from conventional to organic management. However, it is worth to be noticed the lacking of consensus concerning how land management affects separately richness and abundance of spiders (see Prieto-Benítez & Méndez, 2011). Literature mostly focuses differences in the biodiversity of spiders across three classical land managements: organic, integrated and conventional regardless the internal spatial structure of each crop. However, the structure of crops (e.g., vegetation and soil structure and local activities such as grazing) may be more influent in how spider communities differ among managements than the farming system as a global point of view (Bengtsson et al., 2005).

In Spain, Cárdenas & Barrientos (2011) identified 32 families, 115 genera and 142 species in samples collected in Andalusian olive groves. In terms of guilds, ground runners, stalkers, tangled web builders and space web builders were the most abundant on the ground (Cárdenas et al., 2015). In the canopy, Morris et al., (1999) cited Thomisidae and Theridiidae as the families that presented high species richness.

Cárdenas (2008) studied the use of spiders as bioindicators in the olive grove, suggesting that species composition does not vary between different agronomic practices, but between different landscape scales. Castro et al., (1996) reported that bare soil presented lower abundance of spiders on the ground near olive trees. This is supported by the results obtained by Cárdenas et al. (2015) who found a negative effect on the abundance and diversity of spiders induced by plowing, nevertheless it could be reduced by the presence of hedge vegetation. On the contrary, an intermediate disturbance level in olive groves managed according the integrated protection system may lead to an increase in the diversity of spiders (Cárdenas et al., 2006) and Cárdenas et al. (2005) found that *Thyene imperialis* Rossi, 1846 (Salticidae) and *Loxosceles rufescens* Dufour, 1820 (Sicariidae) may be favored by cereal cover crops.

In Italy, Thaler & Zapparoli (1993) reported Linyphiidae, Erigonidae, Lycosidae, Dysderidae, Gnaphosidae and Theridiidae as the dominant spider families for an assemblage of 70 species belonging to 18 families on the ground around olives and Picchi et al. (2016) showed that the distance towards the field edge and the agricultural management affected the community of spiders in the olive grove. The abundance and species richness of spiders as well as the abundance of sheet web spiders were lower in conventional orchards than in organic orchards and Linyphiidae density decreased with the increase of the area shrublands surrounding olive groves.

In Southern Portugal, Sousa (2003) found higher diversity of spiders in organic than in intensively managed groves both in terms of abundance and richness and reported 31 families, 104 genera and 144 species being the most abundant *Hogna radiata* Latreille, 1817 (Lycosidae), *Zodarion josefinae* Bosmans, 1994 (Zodariidae), *Pardosa proxima* (C. L. Koch, 1847) (Lycosidae), *Nomisia exornata* (C. L. Koch, 1839) (Gnaphosidae) and *Callilepis concolor* Simon, 1914. Similarly, in northeastern Portugal, Benhadi-Marín et al., (data not published) found that *N. exornata* (Gnaphosidae), *Zodarion alacre* (Simon, 1870) (Zodariidae) and *Alopecosa albofasciata* (Brullé, 1832) (Lycosidae) were the most abundant species, and grouped the whole community in nine guilds namely: foliage runners, ground runners, stalkers, ambushers, sheet web builders, wandering sheet/tangle weavers, orb weavers, space web builders and specialists. According to the dominant families, ground runners and specialists were the two most representative functional groups and were present in all managements. In this case, foliage runners and orb weavers were exclusive of groves with spontaneous vegetation covering the soil. The remaining guilds were common to all management systems except the sheet web builders that were not present in groves treated with herbicides.

In Iran, Ghavami (2006) collected 18 families, distributed in 56 genera and 18 species in olive orchards and reported *Frontinellina frutetorum* (C. L. Koch, 1834) (Linyphiidae), *T. imperialis* and *Salticus scenicus* (Clerck, 1757) (Salticidae) as the dominant species and, in general, spiders represented almost half (42.8%) of the total abundance of predators collected.

Considering the application of products for pest control, in Spain, Pascual et al. (2010) referred Philodromidae family as being one of the most affected taxa among the community of natural enemies after spraying kaolin on olive trees. In Italy, Scalercio et al. (2009) did not find effects on the community of spiders after spraying rotenone, azadirachtin and copper oxychloride on olive trees (pesticides allowed in organic farming). Regarding spiders as natural enemies, Picchi et al. (2016) found that the density of *Bactrocera oleae* (Rossi, 1790), the key pest of olives in the Mediterranean region, were negatively correlated with the abundance of cursorial and sheet web spiders suggesting that spiders could have suppressed the population of *B. oleae*.

In summary, the conservation of a diversified surrounding landscape as well as the ecological infrastructures growing within the grove (e.g., vegetation strips, small bushes and stonewalls) can be seen as part of a strategy to enhance spider functional biodiversity towards a biological control strategy of olive tree pests. In addition, environmental friendly practices such as organic field crop management can be a way of achieving sustainability and giving an add-value to olive products especially in regions where the landscape is typically heterogeneous, consisting of small mosaics of olive groves, other crops and natural vegetation.

CONCLUSION

Spiders are among the most abundant and widespread predators all around the world, being considered potential biological control agents of pests in agroecosystems. However, alternative non-prey foods and shelter within and between crops are necessary in order to improve their presence and facilitate their permanence in the agricultural landscape. Such resources can be provided by maintaining a heterogeneous agricultural landscape and ecological infrastructures such as flower strips (providing pollen and nectar) and stones and field edges (providing shelter and overwintering sites). In general, common agricultural practices such as pesticide application and soil disturbance caused by tilling usually negatively affect the assemblage of spiders in four ways: (1) by reducing the available habitat at different scales (e.g., cyclic soil disturbance and loss of vegetation complexity due to herbicide application), (2) by reducing the available prey populations (e.g., insecticide application), (3) by reducing sources of alternative non-prey foods (e.g., grass cutting and weed control) and by (4) direct mortality due to intoxication with synthetic compounds (e.g., pesticide applications). Environmentally friendly practices such as organic farming can effectively improve the population of spiders in arable lands thus enhancing the role of spiders as natural enemies specially when crops are integrated in a diversified landscape. In addition, novel bioinsecticides based on spider venom proteins have appeared in the market during the last years. This technology has succeeded in developing pesticides with less lethal and sublethal effects on beneficial fauna and represents a new generation of bioinsecticides allowed in organic agriculture.

ACKNOWLEDGMENTS

The authors are grateful to the Portuguese Foundation of Science and Technology (FCT) for financial support through the projects EXCL/AGR-PRO/0591/2012 “Olive crop protection in sustainable production under global climatic changes: linking ecological infrastructures to ecosystem functions.” J. Benhadi-Marín is grateful to FCT for the PhD studentship (SFRH/BD/97248/2013); S. A. P. Santos is grateful to FCT for the Post-doctoral fellowship (SFRH/BPD/102277/2014). This chapter is part of J. Benhadi-Marín PhD Thesis.

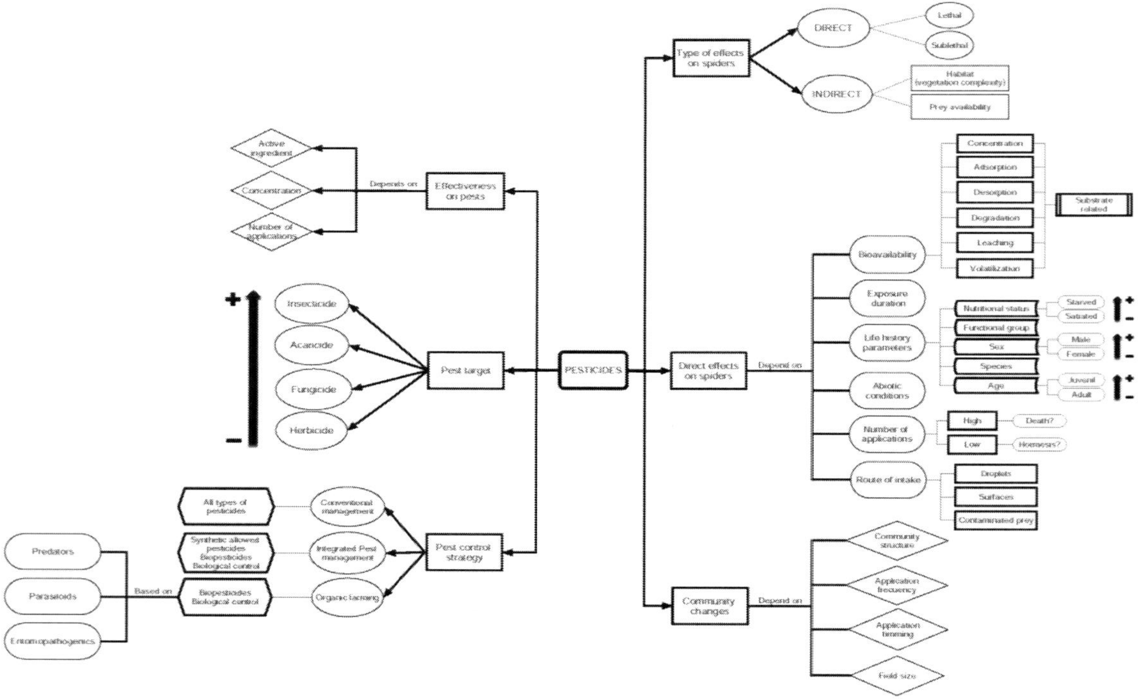

Annex 1. Conceptual overview on how pesticides affect spiders. Arrows indicate a toxicity gradient, from low (-) to high (+), of different pesticides on spiders (left chart side) and how different life history traits are affected by them (right chart side). Based on Pékar (2012).

REFERENCES

Altieri, M. A. & Schmidt, L. L. (1986). Cover crops affect insect and spider populations in apple orchards. *Californian Agriculture, 40*, 15–17.

Balogh, J. & Loksa, J. (1948). Quantitative-biosoziologische untersuchung der Arthropodenwelt ungarischer sandgebiete. *Archiva Biologica Hungarica, 2*, 65–100.

Barrientos, J. A. (1988). Araneae. In: J. A. Barrientos, *Bases Para un Curso Práctico de Entomología.* (pp. 115–141). Salamanca: Asociacion Española de Entomología.

Batáry, P., Holzschuh, A., Orci, K. M., Samu, F. & Tscharntke, T. (2012). Responses of plant, insect and spider biodiversity to local and landscape scale management intensity in cereal crops and grasslands. *Agriculture, Ecosystems and Environment, 146*, 130–136.

Bell, J. R., Johnson, P. J., Hambler, C., Haughton, A. J., Smith, H., Feber, R. E., Tattersall, F. H., Hart, B. H., Manley, W. & Macdonald, D. W. (2002). Manipulating the abundance of *Lepthyphantes tenuis* (Araneae: Linyphiidae) by field margin management. *Agriculture, Ecosystems and Environment, 93*, 295–304.

Benamú, M. A., Schneider, M. I. & Sánchez, N. E. (2010). Effects of the herbicide glyphosate on biological attributes of *Alpaida veniliae* (Araneae, Araneidae), in laboratory. *Chemosphere, 78*, 871–876.

Bengtsson, J., Ahnström, J., Weibull, A.-C. (2005). The effects of organic agriculture on biodiversity and abundance: a meta-analysis. *Journal of Applied Ecology, 42*, 261–269.

Benhadi-Marín, J., Pereira, J. A, Barrientos, J. A., Bento, A. & Santos, S. A. P. (2013). Araneae communities associated with the canopies of chestnut trees in the northeastern part of Portugal: The influence of soil management practices. *European Journal of Entomolology, 110*, 501–508.

Benhadi-Marín, J., Pereira, J. A. & Santos, S. A. P. 2016. Effects of kaolin particle films on the life span of an orb-weaver spider. *Chemosphere, 144*: 918–924.

Bera, S. K., Trivedi, A. & Sharma, C. (2002). Trapped pollen and spores from spider webs of Lucknow environs. *Current Science, 83* 1580–1585.

Berland, L. (1938). *Les Araignées*. Paris: Editions Stock.

Birkhofer, K., Entling, M. H. & Yael, L. (2013). Agroecology: trait composition, spatial relationships, trophic interactions. In: D. Penney (Ed.) *Spider Research in the 21st Century: Trends & Perspectives*, (pp. 200–228). Manchester: Siri Scientific Press.

Bishop, A. L. & Blood, P. R. B. (1981). Interactions between natural populations of spiders and pests in cotton and their importance to cotton production in southeastern Queensland. *General & Applied Entomology, 13*, 98–104.

Blagbrough, I. S., Brackie, P. T. H., Bruce, M., Bycroft, B. W., Mather, A. J., Millington, S., Sudan, H. L. & Usherwood, P. N. R. (1992). Arthropod toxins as leads for novel insecticides: an assessment of polyamine amides as glutamate antagonists. *Toxicon, 30*, 303–322.

Blumberg, A. Y. & Crossley, D. A. Jr. (1983). Comparison of soil surface arthropod populations in conventional tillage, no-tillage and old field systems. *Agro-Ecosystems, 8*, 247–253.

Bond, J. E., Garrison, N. L., Hamilton, C. A., Godwin, R. L., Hedin, M. & Garson, I. (2014). Phylogenomics resolves a spider backbone phylogeny and rejects a prevailing paradigm for orb web evolution. *Current Biology, 24*, 1765–1771.

Bristowe, W. S. (1958). *The world of spiders*. London: Collins.

Brusca, R. C. & Brusca, G. J. (2005). *Invertebrados*. Madrid: McGraw-Hill, Interamericana de España.

Butt, A. & Sherawat, S. M. (2012). Effect of different agricultural practices on spiders and their prey populations in small wheat fields. *Acta Agriculturae Scandinavica, Section B - Soil & Plant Science, 62*, 374–382.

Caraco, T. & Gillespie, R. G. (1986). Risk-sensitivity: foraging mode in an ambush predator. *Ecology, 67,* 1180–1185.

Cárdenas, M. (2008). *Análisis de la Actividad Ecológica de las Arañas en el Agroecosistema del Olivar*. PhD Thesis. Granada: Universidad de Granada.

Cárdenas, M. & Barrientos, J. A. (2011). Spiders from Andalusian olive groves (Arachnida; Araneae). Faunistic aspects. *Zoologica Baetica, 22*: 99–136.

Cárdenas, M., Barrientos, J. A., García, P., Pascual, F. & Campos, M. (2005). Effect of cereal cover crops on Araneae population in olive orchard, In: A. Kalaitzaki, (Ed.) *Proceedings of the 2nd European Meeting of the IOBC/WPRS Working Group "Integrated Protection of Olive Crops,"* 26-28 October 2005. Firenze, Italy. pp. 207.

Cárdenas, M., Pascual, F., Campos, M. & Pékar, S. (2015). The spider assemblage of olive groves under three management systems. *Environmental Entomology, 44*: 509–518.

Cardoso, P., Pekár, S., Jocqué, R. & Coddington, J. A. (2011). Global patterns of guild composition and functional diversity of spiders. *PLoS ONE, 6*, e21710.

Castro, J., Campos, P. & Pastor, M. (1996). Influencia de los sistemas de cultivo empleados en olivar y girasol sobre la composición de la fauna de artrópodos en el suelo. *Boletín de Sanidad Vegetal, Plagas, 22*, 557–570.

Chapman, E. G., Schmidt, J. M., Welch, K. D. & Harwood, J. D. (2013). Molecular evidence for dietary selectivity and pest suppression potential in an epigeal spider community in winter wheat. *Biological Control, 65*, 72–86.

Chatterjee, S., Isaia, M. & Venturino, E. (2009). Spiders as biological controllers in the agroecosystem. *Journal of Theoretical Biology, 258*, 352–362.

Chen, X., Chen, Y., Wu, L., Peng, Y., Chen, J. & Liu, F. (2010). A survey of nectar feeding by spiders in three different habitats. *Bulletin of Insectology, 63*, 203–208.

Cherrett, J. M. (1964). The distribution of spiders on the Moor House National Reserve, Westmorland. *Journal of Animal Ecology, 33*, 27–48.

Chiverton, P. A. (1986). Predator density manipulation and its effect on populations of *Rhopalosiphum padi* (Hom.: Aphididae) in spring barley. *Annals of Applied Biology, 109*, 49–60.

Comstock, J. H. (1940). The Spider Book. (revised, edited by W. J. Gertsch). Ithaca: Cornell University Press.

de Bello, F., Lavorel, S., Díaz, S., Harrington, R., Cornelissen, J. H. C., Bardgett, R. D., Berg, M. P., Cipriotti, P., Feld, C. K., Hering, D., da Silva, P. M., Potts, S. G., Sandin, L., Sousa, J. P., Storkey, J., Wardle, D. A. & Harrison, P. A. (2010). Towards an assessment of multiple ecosystem processes and services via functional traits. *Biodiversity Conservation, 19*: 2873–2893.

Decourtye, A., Devillers, J., Cluzeau, S., Charreton, M. & Pham-Delègue, M.-H. (2004a). Effects of imidacloprid and deltamethrin on associative learning in honeybees under semi-field and laboratory conditions. *Ecotoxicology and Environmental Safety, 57*, 410–419.

Decourtye, A., Armengaud, C., Renou, M., Devillers, J., Cluzeau, S., Gauthier, M. & Pham-Delègue, M.-H. (2004b). Imidacloprid impairs memory and brain metabolism in the honeybee (*Apis mellifera* L.). *Pesticide Biochemistry and Physiology, 78*, 83–92.

Desneux, N., Decourtye, A. & Delpuech, J.-M. (2007). The sublethal effects of pesticides on beneficial arthropods. *Annual Review of Entomology, 52*, 81–106.

Dicks, L. V., Ashpole, J. E., Dänhardt, J., James, K., Jönsson, A., Randall, N., Showler, D. A., Smith, R. K., Turpie, S., Williams, D. & Sutherland, W. J. (2013). Farmland Conservation: Evidence for the Effects of Interventions in Northern and Western Europe. Exeter: Pelagic Publishing.

Diehl, E., Mader, V. L., Wolters, V. & Birkhofer, K. (2013). Management intensity and vegetation complexity affect web-building spiders and their prey. *Oecologia, 173*, 579–589.

Dunlop, J. A. & Penney, D. (2011). Order Araneae Clerck, 1757. In: Z.-Q. Zhang (Ed.) *Animal biodiversity: An outline of higher-level classification and survey of taxonomic richness*. Zootaxa, 3148, 149–153.

Eggs, B. & Sanders, D. (2013). Herbivory in spiders: the importance of pollen for orb-weavers. *PLoS ONE, 8*, e82637.

Ehler, L. (1998). Conservation biological control: Past, Present, and Future. In: P. Barbosa (Ed.), *Conservation Biological Control*. (pp 1–8). San Diego: Academic Press.

El-Wakeil, N., Gaafar, N., Sallam, A. & Volkmar, C. (2013). Side effects of insecticides on natural enemies and possibility of their integration in plant protection strategies. In: S. Trdan (Ed.) *Insecticides: Development of Safer and More Effective Technologies Agricultural and Biological Sciences*. (pp. 1–56). InTech. Doi: 10. 5772/54199.

Enders, F. (1974). Vertical stratification in orb-web spiders (Araneidae, Araneae) and a consideration of other means of coexistence. *Ecology, 55*, 317–328.

Estevão, A., Santos, S. A. P., Silva, A., Gonçalves, C., Pereira, J. A. & Patanita, M. I. (2012). Effect of the plant protection systems on soil arthropods in olive groves from Alentejo region (South-East of Portugal). Proceedings of Working Group "Integrated Protection of Olive Crops," Jerusalem, Israel. 15-20 May 2011. *IOBC/WPRS Bulletin, 79*, 173–178.

European Commission. (2011). Communication from the Commission to the European Parliament, the Council, the Economic and Social Committee and the Committee of the Regions: Our life insurance, our natural capital: an EU biodiversity strategy to 2020. COM (2011) 244 final. Brussels. http://www.ec.europa.eu/environment/nature/biodiversity/comm2006/2020.htm. Accessed [4/05/2016].

Evans, S. C., Shaw, E. M. & Rypstra, A. L. (2010). Exposure to a glyphosate-based herbicide affects agrobiont predatory arthropod behaviour and long-term survival. *Ecotoxicology, 19*, 1249–1257.

Fisher, R. A., Corbert, A. S. & Williams, C. B. (1943). The relation between the number of species and the number of individuals in a random sample of an animal population. *Journal of Animal Ecology, 12*, 42–58.

Flynn, D. F. B., Gogol-Prokurat, M., Nogeire, T., Molinari, N., Richers, B. T., Lin, B. B., Simpson, N., Mayfield, M. M. & DeClerck, F. (2009). Loss of functional diversity under land use intensification across multiple taxa. *Ecology Letters, 12,* 22–33.

Foelix, R. F. (1996). *Biology of Spiders*. Oxford: Oxford University Press.

Geiger, F., Bengtsson, J., Berendse, F., Weisser, W. W., Emmerson, M., Morales, M. B., Ceryngier, P., Liira, J., Tscharntke, T., Winqvist, C., Eggers, S., Bommarco, R., Pärt, T., Bretagnolle, V., Plantegenest, M., Clement, L. W., Dennis, C., Palmer, C., Oñate, J. J., Guerrero, I., Hawro, V., Aavik, T., Thies, C., Flohre, A., Hänke, S. , Fischer, C., Goedhart, P. W. , Inchausti, P. (2011). Persistent negative effects of pesticides on biodiversity and biological control potential on European farmland. *Basic and Applied Ecology, 11*, 97–105.

Gertsch, W. J. (1979). *American Spiders*. New York: Van Nostrand Reinhold.

Ghavami, S. (2006). Abundance of spiders (Arachnida: Araneae) in olive orchards in northern part of Iran. *Pakistan Journal of Biological Sciences, 9*, 795–799.

Ghavami, S. (2008). The Potential of predatory spiders as biological control agents of cotton pests in Tehran provinces of Iran. *Asian Journal of Experimental Sciences, 22*, 303–306.

Glenn, D. M. & Puterka, G. J. (2005). Particle films: a new technology for agriculture. In: J. Janick, (ed.). *Horticultural Reviews*, (pp. 1–44). Oxford:. John Wiley & Sons Inc.

Glenn, D. M., Outerka, G., Vanderzwet, T., Byers, R. & Feldhake, C. (1999). Hydrophobic particle films: a new paradigm for suppression of arthropod pests and plant diseases. *Journal of Economic Entomology, 92*, 759–771.

Greenstone, M. H. (1999). Spider predation: how and why we study it. *The Journal of Arachnology, 27,* 333–342.

Greenstone, M. H. & Shufran, K. A. (2003). Spider predation: species specific identification of gut contents by polymerase chain reaction. *Journal of Arachnology, 31,* 131–134.

Gregory, B. M. (1989). Field observations of *Gasteracantha cancriformis* (Araneae, Araneidae) in a Florida mangrove stand. *The Journal of Arachnology, 17*, 119–120.

Harrington, R., Anton, C., Dawson, T. P., de Bello, F., Feld, C. K., Haslett, J.R., Kluvánkova-Oravská, T., Kontogianni, A., Lavorel, S., Luck, G. W., Rounsevell, M. D. A., Samways, M. J., Settele, J., Skourtos, M., Spangenberg, J. H., Vandewalle, M., Zobel, M., Harrison, P. A. (2010). Ecosystem services and biodiversity conservation: concepts and a glossary. *Biodiversity Conservation, 19,* 2773–2790.

Harwood, J. D., Phillips, S. W., Lello, J., Sunderland, K. D., Glen, D. M., Bruford, M. W., Harper, G. L. & Symondson, W. O. C. (2009). Invertebrate biodiversity affects predator fitness and hence potential to control pests in crops. *Biological Control, 51*, 499–506.

Hawkes, N. J., Janes, R. W., Hemingway, J. & Vontas, J. (2005). Detection of resistance-associated point mutations of organophosphate-insensitive acetylcholinesterase in the olive fruit fly, *Bactrocera oleae* (Gmelin). *Pest Biochemistry and Physiology, 81*, 154–163.

Hayashi, M., Bakkali, M., Hyde, A. & Goodacre, S. L. (2015). Sail or sink: novel behavioural adaptations on water in aerially dispersing species. *Evolutionary Biology, 15*, 118.

Herzig, V., Wood, D. L. A., Newell, F., Chaumeil, P.-A., Kaas, Q., Binford, G. J., Nicholson, G. M., Gorse, D. & King, G. F. (2011). ArachnoServer 2.0, an updated online resource for spider toxin sequences and structures. Nucleic Acids Research 39, D653-D657. Available at http://www. arachnoserver.org Accessed [29/04/2016],

Hodgson, E. (1984). Development of safer insecticides. *Drug Metabolism Reviews, 15*, 881–895.

Holland J. M. & Reynolds C. R. (2003). The impact of soil cultivation on arthropod (Coleoptera and Araneae) emergence on arable land. *Pedobiologia, 47*, 181–191.

Hubert, M. (1979). *Les araignées*. Paris: Société Nouvelle des Editions Boubee.

Jackson, R. R., Pollard, S. D., Nelson, X. J., Edwards, G. B. & Barrion, A. T. (2001). Jumping spiders (Araneae: Salticidae) that feed on nectar. *Journal of Zoology, 255*, 25–29.

Jeanneret, P., Schüpbach, B. & Luka, H. (2003). Quantifying the impact of landscape and habitat features on biodiversity in cultivated landscapes. *Agriculture, Ecosystems and Environment, 98*, 311–320.

Jiménez-Valverde, A. & Lobo, J. M. (2007). Determinants of local spider (Araneidae and Thomisidae) species richness on a regional scale: climate and altitude vs. habitat structure. *Ecological Entomology, 32*, 113–122.

Jones, S. E. (1941). Influence of temperature and humidity on the life history of the spider *Agelena naevia* Walckenaer. *Annals of the Entomological Society of America, 34*, 557–571.

King, G. F. & Hardy, M. C. (2013). Spider-venom peptides: structure, pharmacology and potential for control of insect pests. *Annual Review of Entomology, 58,* 475–496.

Korenko, S., Niedobová, J., Kolárová, M., Hamouzová, K., Kysilková, K. & Michalko, R. (2016). The effect of eight common herbicides on the predatory activity of the agrobiont spider *Pardosa agrestis. BioControl.* Doi: 10.1007/s10526-016-9729-0.

Kuntner M. & Coddington J. A. (2009). Discovery of the largest orb weaving spider species: the evolution of gigantism in *Nephila. PLoS ONE, 4*, e7516.

Kuusk, A.-K., Cassel-Lundhagen, A., Kvarnheden, A. & Ekbom, B. (2008). Tracking aphid predation by lycosid spiders in spring-sown cereals using PCR-based gut-content analysis. *Basic and Applied Ecology, 9*, 718–725.

Lafage, D. & Pétillon, J. (2016). Relative importance of management and natural flooding on spider, carabid and plant assemblages in extensively used grasslands along the Loire. *Basic and Applied Ecology.* Doi: 10.1016/j.baae.2016.04.002.

László, M., Csaba, N. & Markó, V. (2015). Canopy dwelling hunting spider assemblages on apple trees and their ability to control pests. *Növényvédelem, 51*, 409–416.

Lavorel, S., McIntyre, S., Landsber, J. & Forbes, T. D. A. (1997). Plant functional classification: from general groups to specific groups based on response to disturbance. *Trends in Ecology & Evolution, 12*, 474–478.

Lawes, M. J., Kotze, D. J., Bourquin, S. L. & Morris, C. (2005). Epigeic invertebrates as potential ecological indicators of afromontane forest condition in South Africa. *Biotropica, 37*, 109–118.

Leccia, F., Kysilkova, K., Kolarova, M., Hamouzova, K., Liznarova, E. & Korenko, S. (2016). Disruption of the chemical communication of the European agrobiont ground-dwelling spider *Pardosa agrestis* by pesticides. *Journal of Applied Entomology*. doi: 10.1111/jen.12288.

Longcore, T. (2003). Terrestrial arthropods as indicators of ecological restoration success in coastal sage scrub (California, USA). *Restoration Ecology, 11*, 397–409.

Luczak, J. (1960). Rozmieszczenie pietrowe pajakow w lesie. *Ekologia Polska (Seria B), 6*, 39–50.

Luczak, J. (1963). Differences in the structure of communities of web spiders in one type of environment (young pine forest). *Ekologia Polska (Seria A), 11*, 159–221.

Luczak, J. (1966). The distribution of wandering spiders in different layers of the environment as a result of interspecies competition. *Ekologia Polska (Seria A), 14*, 233–244.

Maloney, D., Drummond, F. A. & Alford, R. (2003). *Spider Predation in Agroecosystems: Can Spiders Effectively Control Pest Populations?* Technical Bulletin 190. Orono: Maine Agricultural and Forest Experiment Station, University of Main.

Marc, P., Canard, A. & Ysnel, F. (1999). Spiders (Araneae) useful for pest limitation and bioindication. *Agriculture, Ecosystems and Environment, 74*, 229–273.

Markó, V., Bogya, S., Kondorosy, E. & Blommers, L. H. M. (2010). Side effects of kaolin particle films on apple orchard bug, beetle and spider communities. *International Journal of Pest Management, 56*, 189–199.

Martins, A., Linhares, I., Raimundo, F., Coutinho, J. P., Gomes-Laranjo, J., Borges, O. & Sousa, V. (2005). The importance of deep soil layers to supply water to agro-forestry systems: a case study of a mature chestnut orchard in Northern Portugal. *Acta Horticulturae, 693*, 663–670.

Michalková, V. & Pekár, S. (2009). How glyphosate altered the behaviour of agrobiont spiders (Araneae: Lycosidae) and beetles (Coleoptera: Carabidae). *Biological Control, 51*, 444–449.

Monzó, C., Mollá, Ó., Castañera, P. & Urbaneja, A. (2009). Activity-density of *Pardosa cribata* in Spanish citrus orchards and its predatory capacity on *Ceratitis capitata* and *Myzus persicae*. *BioControl, 54*, 393–402.

Monzó, C., Sabater-Muñoz, B., Urbaneja, A. & Castanera, P. (2010). Tracking medfly predation by the wolf spider, *Pardosa cribata* Simon, in citrus orchards using PCR-based gut-content analysis. *Bulletin of Entomological Research, 100*, 145–152.

Moreno, C. E. (2001). *Métodos para Medir la Biodiversidad.* Zaragoza: Sociedad Entomología Aplicada.

Morris, T., Symondson, W. O. C., Kidd, N. A. C. & Campos, M. (1999). Las arañas y su incidencia sobre *Prays oleae* en el olivar. *Boletín de Sanidad Vegetal, Plagas, 25*, 475–489.

Morse, D. H. (2007). *Predator Upon a Flower, Life History and Fitness in a Crab Spider.* Cambridge: Harvard University Press.

Mulhauser, B. (1990). La bioindication?... Et si nous reparlions des araignées? In: M. L. Célérier, J. Heurtault & C. Rollard, C. (Eds.), Proc. 12ème Colloque Européen d'Arachnologie. *Bulletin de la Société Européenne d'Arachnologie, 1*, 266–272.

Murdoch, W. W., Chesson, J. & Chesson, P. L. (1985). Biological control in theory and practice. *The American Naturalist, 125*, 344–366.

Nakamura, M. & Nakamura, K. (1977). Population dynamics of the chestnut gall wasp, *Dryocosmus kuriphilus* Yasumatsu (Hymenoptera: Cynipidae) V. Estimation of the effect of predation by spiders on the mortality of imaginal wasps based on the precipitin test. *Oecologia, 27*, 97–116.

Nakasu, E. Y. T, Williamson, S. M., Edwards, M. G., Fitches, E. C., Gatehouse, J. A., Wright, G. A. & Gatehouse, A. M. R. (2014). Novel biopesticide based on a spider venom peptide shows no adverse effects on honeybees. *Proceedings of the Royal Society B: Biological Sciences, 281*, 20140619.

Nentwig, W. (1993). Spider venoms are not suitable insecticides. *Toxicon, 31*, 233–236.

Nentwig, W. & Wissel, C. (1986). A comparison of prey lengths among spiders. *Oecologia, 68*, 595–600.

Nentwig, W., Frank, T. & Lethmayer, C. (1998). Sown weed strips: artificial ecological compensation areas as an important tool in conservation biological control. In: P. Barbosa (Ed.), *Conservation Biological Control.* (pp. 133–153). San Diego: Academic Press.

Nieto, J. M. & Mier, M. P. (1985). *Tratado de Entomología.* Barcelona: Ediciones Omega.

Nyffeler, M. (1982). *Field studies on the ecological role of spiders as insect predators in agroecosystems (abandoned grassland, meadows, and cereal fields).* PhD. Thesis. Zurich Swiss Federal Institute of Technology.

Nyffeler, M. & Sunderland, K. D. (2003). Composition, abundance and pest control potential of spider communities in agroecosystems: a comparison of European and US studies. *Agriculture, Ecosystems and Environment, 95*, 579–612.

Nyffeler, M., Olson, E. J. & Symondson, W. O. C. (2016). Plant-eating by spiders. *Journal of Arachnology, 44*, 15–27.

Öberg, S. (2009). Influence of landscape structure and farming practice on body condition and fecundity of wolf spiders. *Basic and Applied Ecology, 10*, 614–621.

Öberg, S., Mayr, S. & Dauber, J. (2008). Landscape effects on recolonisation patterns of spiders in arable fields. *Agriculture, Ecosystems and Environment, 123*, 211–218.

Oelbermann, K. & Scheu, S. (2009). Control of aphids on wheat by generalist predators: effects of predator density and the presence of alternative prey. *Entomologia Experimentalis et Applicata, 132*, 225–231.

Palmer, M. J., Moffat, C., Saranzewa, N., Harvey, J., Wright, G. A. & Connolly, C. N. (2013). Cholinergic pesticides cause mushroom body neuronal inactivation in honeybees. *Nature Communications, 4*, 1634

Pascual, S., Cobos, G., Seris, E. & González-Núñez, M. (2010). Effects of processed kaolin on pests and non-target arthropods in a Spanish olive grove. *Journal of Pest Science, 83*, 121–133.

Pekár, S. (2012). Spiders (Araneae) in the pesticide world: an ecotoxicological review. *Pest Management Science, 68*, 1438–1446.

Peterson, J. A., Romero, S. A. & Harwood, J. D. (2010). Pollen interception by linyphiid spiders in a corn agroecosystem: implications for dietary diversification and risk-assessment. *Arthropod-Plant Interactions, 4*, 207–217.

Pétillon, J., Montaigne, W. & Renault, D. (2009). Hypoxic coma as a strategy to survive inundation in a salt-marsh inhabiting spider. *Biology Letters, 5*, 442–4.

Pfister, S. C., Schäfer, R. B., Schirmel, J. & Entling, M. H. (2015). Effects of hedgerows and riparian margins on aerial web-building spiders in cereal fields. *Journal of Arachnology, 43*, 400–405.

Pianka, E. R. (1966). Convexity, desert lizards, and spatial heterogeneity. *Ecology, 47*, 1055–1059.

Picchi, M. S., Bocci, G., Petacchi, R. & Entling, M. H. (2016). Effects of local and landscape factors on spiders and olive fruit flies *Agriculture, Ecosystems and Environment, 222*, 138–147.

Platnick, N. I. (1971). The evolution of courtship behaviour in spiders. *Bulletin of the British Arachnological Society, 2*, 40–47.

Pollard, S. D., Beck, M. W. & Dodson, G. N. (1995). Why do male crab spiders drink nectar? *Animal Behaviour, 49*, 1443–1448.

Post, W. M. & Riechert, S. E. (1977). Initial investigation into the structure of spider communities. *Journal of Animal Ecology, 46*, 729–749.

Prieto-Benítez, S. & Méndez, M. (2011). Effects of land management on the abundance and richness of spiders (Araneae): A meta-analysis. *Biological Conservation, 144*, 683–691.

Pywell, R. F., James, K. L., Herbert, I., Meek, W. R., Carvell, C., Bell, D. & Sparks, T. H. (2005). Determinants of overwintering habitat quality for beetles and spiders on arable farmland. *Biological Conservation, 123*, 79–90.

Quicke, D. L. J. & Usherwood, P. N. R. (1990). Spider toxins as lead structures for novel pesticides. In: E. Hodgson & R. J. Kuhr (Eds.) *Safer Insecticides: Development and Use.* (pp 385–452). New York: Marcell Dekker Inc.

Riechert, S. E. (1999). The hows and whys of successful pest suppression by spiders: insights from case studies. *The Journal of Arachnology, 27*, 387–396.

Rierchert, E. S. & Lockley, T. (1984). Spiders as biological control agents. *Annual Review of Entomology, 29*, 299–320.

Riechert, S. E. & Maupin, J. L. (1998). Spider effects on prey: Tests for superfluous killing in five web-builders. In: P. A. Selden (Ed.). *Proceedings of the 17th European Colloquium of Arachnology,* Burnham Beeches (pp. 203–210). Edinburgh: British Arachnological Society.

Roberts, M. J. (1996). *Spiders of Britain & Northern Europe.* London: Collins Publishers.

Růžička, V. (1990). Size structure of epigeic communities of spiders, carabids and staphylinids (Araneae; Coleoptera: Carabidae, Staphylinidae). *Acta Universitatis Carolinae - Biologica, 34*, 263–274.

Rypstra, A. L., Carter, P. E., Balfour, R. A., & Marshall, S. D. (1999). Architectural features of agricultural habitats and their impact on the spider inhabitants. *The Journal of Arachnology, 27*, 371–377.

Sackett, T. E., Buddle, C. M. & Vincent, C. (2007). Effects of kaolin on the composition of generalist predator assemblages and parasitism of *Choristoneura rosaceana* (Lep., Tortricidae) in apple orchards. *Journal of Applied Entomology, 131*, 478–485.

Sackett, T. E., Buddle, C. M. & Vincent, C. (2009). Dynamics of spider colonization of apple orchards from adjacent deciduous forest. *Agriculture, Ecosystems and Environment, 129*, 144–148.

Sage, R. D. (1982). Wet and dry-weight estimates of insects and spiders based on length. *The American Midland Naturalist, 108*, 407–411.

Samu, F. & Biro, Z. (1993). Functional response, multiple feeding and wasteful killing in a wolf spider (Araneae: Lycosidae). *European Journal of Entomology, 90*, 471–476.

Samu, F. & Szinetár, C. (2002). On the nature of agrobiont spiders. *The Journal of Arachnology, 30*, 389–402.

Samu, F., Sunderland, K. D. & Szinetár, C. (1999). Scale-dependent dispersal and distribution patterns of spiders in agricultural systems: a review. *Journal of Arachnology, 27*, 325–332.

Samu, F., Sunderland, K. D., Topping, C. J. & Fenlon, J. S. (1996). A spider population in flux: selection and abandonment of artificial web-sites and the importance of intraspecific interactions in *Lepthyphantes tenuis* (Araneae: Linyphiidae) in wheat. *Oecologia, 106*, 228–239.

Scalercio, S., Belfiore, T., Noce, M. E., Vizzarri, V. & Iannotta, N. (2009). The impact of compounds allowed in organic farming on the above-ground arthropods of the olive ecosystem. *Bulletin of Insectology, 62*, 137–141.

Schaffers, A. P., Raemakers, I. P., Sykora, K. V. & Ter Braak, C. J. (2008). Arthropod assemblages are best predicted by plant species composition. *Ecology, 89*, 782–794.

Schmidt, M. H. & Tscharntke, T. (2005). Landscape context of sheetweb spider (Araneae: Linyphiidae) abundance in cereal fields. *Journal of Biogeography, 32*, 467– 473.

Schmidt, M. H., Thies, C., Nentwig, W. & Tscharntke, T. (2008). Contrasting responses of arable spiders to the landscape matrix at different spatial scales. *Journal of Biogeography, 35*, 157–166.

Schoener, T. W. (1971). Theory of feeding strategies. *Annual Review of Ecology and Systematics, 2*, 369–404.

Shivakumar, M. S. & Kumar, D. (2010). Biological control potential of male and female *Oxyopes shweta* (Araneae: Oxyopidae) against polyphagous insect pest *Spodoptera litura*. *Journal of Ecobiotechnology, 2*, 20–24.

Showler, A. T. (2002). Effects of kaolin-based particle film application on boll weevil (Coleoptera: Curculionidae) injury to cotton. *Journal of Economic Entomology, 95*, 754–762.

Simberloff, D. & Dayan, T. (1991). The guild concept and the structure of ecological communities. *Annual Review of Ecology, Evolution and Systematics, 22*, 115–143.

Smith, R. B. & Mommsen, T. P. (1984). Pollen feeding in an orb-weaving spider. *Science, 226*, 1330–1332.

Soares, M. E., Pereira, J. A. & Bastos, M. L. (2004). Copper residues in olives after olive tree treatments with three different copper formulations. In: B. Kortsen, M. Bickel, K.-H. Grobecker, Book of Abstracts of the 2nd International IUPAC Symposium Trace Elements in Food. 7–8 October. Brussels.

Sousa, A. (2003). *Diversidade de Aranhas nos diferentes sistemas de cultivo de olival no Alentejo (Portugal)*. MSc Thesis. Beja: Instituto Politécnico de Beja.

Suetsugu, K., Hayamizu, M. & Koike, N. (2014). *Clubiona* spider (Araneae: Clubionidae) visiting flowers of nectariferous orchid *Neottianthe cucullata*. *Entomological Science, 17*, 262–264.

Sunderland, K. (1999). Mechanisms underlying the effects of spiders on pest populations. *The Journal of Arachnology, 27*, 308–316.

Sunderland, K. & Samu, F. (2000). Effects of agricultural diversification on the abundance, distribution, and pest control potential of spiders: a review. *Entomologia Experimentalis et Applicata, 95*, 1–13.

Sunderland, K.. D., Fraser, A. M. & Dixon, A. F. G. (1986). Field and laboratory studies on money spiders (Linyphiidae) as predators of cereal aphids. *Journal of Applied Ecology, 23*, 433–447.

Sunderland, K. D., Axelsen, J. A., Dromph, K., Freier, B., Hemptinne, J.-L., Holst, N. H., Mols, P. J. M., Petersen, M. K., Powell, W., Ruggle, P., Triltsch, H., Winder, L. (1997). Pest control by a community of natural enemies. In: W. Powell (Ed.), *Arthropod Natural Enemies in Arable Lands III. The Individual, the Population and the Community*. Acta Jutlandica, 72: 271–326.

Taylor, R. M. & Pfannenstiel, R. S. (2009). How Dietary Plant Nectar Affects the Survival, Growth, and Fecundity of a Cursorial Spider *Cheiracanthium inclusum* (Araneae: Miturgidae). *Environmental Entomology, 38*, 1379–1386.

Thaler, K. & Zapparoli, M. (1993): Epigeic spiders in an olive-grove in central Italy (Aranae). *Redia, 56*, 307–316.

Thorbek, P. & Bilde, T. (2004). Reduced numbers of generalist arthropod predators after crop management. *Journal of Applied Ecology, 41*, 526–38.

Topping, C. J. (1997). Predicting the effect of landscape heterogeneity on the distribution of spiders in agroecosystems using a population dynamics driven landscape-scale simulation model. *Entomological Research in Organic Agriculture, 1997*, 325–336.

Topping, C. J. (1999). An individual-based model for dispersive spiders in agroecosystems: simulations on the effects of landscape structure. *Journal of Arachnology, 27*, 378–386.

Tous, J., Romero, A., Hermoso, J. F. & Ninot, A. (2011). Mediterranean clonal selections evaluated for modern hedgerow olive oil production in Spain. *California Agriculture, 65*, 34–40.

Tretzel, E. (1955). Intragenerische isolation und interspezifische konkurrenz bei spinnen. *Zeitschrift für Morphologie und Ökologie der Tiere, 44*, 43–162.

Turbull, A. L. (1973). Ecology of true spiders (Araneomorphae). *Annual Review of Entomology, 18*, 305–348.

Turnbull, A. L. (1964). The search for prey by a web-building spider *Achaearanea tepidariorum* (C. L. Koch) (Araneae, Theridiidae). *The Canadian Entomologist, 96*, 568–579.

Uetz, G. W. (1977). Coexistence in a guild of wandering spiders. *Journal of Animal Ecology, 46*: 531–541.

Uetz, G. W. (1992). Foraging Strategies of Spiders. *Tree, 7,* 155–159.

Uetz, G. W., Cady, A. B. & Halaj, J. (2000). Modular habitat refugia enhance generalist predators and lower plant damage in soybeans. *Environmental Entomology, 29*, 383–393.

Uetz, G. W., Halaj, J. & Cady, A. (1999). Guild structure of spiders in mayor crops. *The Journal of Arachnology, 27*, 270–280.

Vestaron (2015). EPA Removes Bee Toxicity Warning Statement From Label of SPEAR™ Bioinsecticide From Vestaron Corporation. 1 pp. Avaliable at www.vestaron.com Accessed [29/04/2016].

Vogelei, A. & Greissl, R. (1989). Survival strategies of the crab spider *Thomisus onustus* Walckenaer 1806 (Chelicerata, Arachnida, Thomisidae). *Oecologia, 80*, 513–515.

Windley, M. J., Herzig, V., Dziemborowicz, S. A., Hardy, M. C., King, G. F. & Nicholson, G. M. (2012). Spider-venom peptides as bioinsecticides. *Toxins, 4*, 191–227.

Wise, D. H. (1993). *Spiders in Ecological Webs*. Cambridge: Cambridge University Press.

World Spider Catalog. (2016). World Spider Catalog. Natural History Museum Bern. Available at http://wsc.nmbe.ch. v.17.0. Accessed [28/04/2016].

Wrinn, K. M., Evans, S. C. & Rypstra, A. L. (2012). Predator cues and an herbicide affect activity and emigration in an agrobiont wolf spider. *Chemosphere, 87*, 390–396.

Wyss, E., Niggli, U. & Nentwig, W. (1995). The impact of spider on aphid populations in a strip-managed apple orchard. *Journal of Applied Entomology, 119*, 473–478.

Young, O. P. & Edwards, G. B. (1990). Spiders in United States field crops and their potential effect on crop pests. *Journal of Arachnology, 18*, 1–27.

Zehnder, G., Gurr, G. M., Kühne, S., Wade, M. R., Wratten, S. D. & Wyss, E. (2007). Arthropod pest management in organic crops. *Annual Review of Entomology, 52*, 57–80.

Zhang, Z.-Q. (2011). Phylum Arthropoda von Siebold, 1848. In: Z.-Q. Zhang (Ed.) *Animal biodiversity: An outline of higher-level classification and survey of taxonomic richness.* Zootaxa, 3148,99-103.

Biographical Sketches

Jacinto Benhadi-Marín

Mountain Research Center (CIMO),
School of Agriculture, Polytechnic Institute of Bragança, Bragança, Portugal
Centre for Functional Ecology, Department of Life Sciences,
University of Coimbra, Coimbra, Portugal.

Education: BSc in Biology (Univerity of León, Spain).

Research and Professional Experience:

Biodiversity and taxonomy of spiders (Araneae).

Study of communities of spiders associated with agro-ecosystems and their potential as biocontrol agents of pests associated to the cultures.

Identification of species of arthropods which can be used as biological indicators in olive crops.

Ethology of agrobiont spiders.

Functional ecology and distribution of spiders within the Iberian Peninsula.

Breeding, rearing and maintaining of spider populations in laboratory.

Data analysis and modeling using R.

Education: BSc in Biology.

Education at other institutions: Currently - PhD student at University of Coimbra (Portugal).

Publications Last Three Years:

Lüscher, G., Ammari, Y., Andriets, A., Angelova, S., Arndorfer, M., Bailey, D., Balázs, K., Bogers, M., Bunce, R. G.H., Choisis, J.-P., Dennis, P., Díaz, M., Dyman, T., Eiter, S., Fjellstad, W., Fraser, M., Friedel, J. K., Garchi, S., Geijzendorffer, I. R., Gomiero, T., González-Bornay, G., Guteva, Y., Herzog, F., Jeanneret, P., Jongman, R. H.G., Kainz, M., Kwikiriza, N., Díaz, M. L. L., Moreno, G., Nicholas-Davies, P., Nkwiine, C., Opio, J., Paoletti, M. G., Podmaniczky, L., Pointereau, P., Pulido, F., Sarthou, J.-P., Schneider, M. K., Sghaier, T., Siebrecht, N., Stoyanova, S., Wolfrum, S., Yashchenko, S., Albrecht, H., Báldi, A., Belényesi, M., Benhadi-Marin, J., Blick, T., Buholzer, S., Centeri, C., Choisis, N., Cuendet, G., De Lange, H. J., Déjean, S., Deltshev, C., Cosín, D. J. D., Dramstad, W., Elek, Z., Engan, G., Evtushenko, K., Falusi, E., Finch, O.-D., Frank, T., Gavinelli, F., Genoud, D., Gillingham, P. K., Grónás, V., Gutiérrez, M., Häusler, W., Heer, X., Hübner, T., Isaia, M., Jerkovich, G., Jesus, J. B., Kakudidi, E., Kelemen, E., Koncz, N., Kovacs, E., Kovács-Hostyánszki, A., Last, L., Ljubomirov, T., Mandery, K., Mayr, J., Mjelde, A., Muster, C., Nascimbene, J., Neumayer, J., Ødegaard, F., Sánchez, F. J. O., Oschatz, M.-L., Papaja-Hülsbergen, S., Paschetta, M., Pavett, M., Pelosi, C., Penksza, K., Pommeresche, R., Popov, V., Radchenko, V., Richner, N., Riedel, S., Scullion, J., Sommaggio, D., Szalkovszki, O., Szerencsits, E., Trigo, D., Vale, J., van Kats, R., Vasilev, A., Whittington, A. E., Wilkes-Allemann, J., Zanetti, T. 2016. Farmland biodiversity and agricultural management on 237 farms in 13 European and 2 African regions. Ecology. doi:10.1890/15-1985.1

Villa, M., Santos, S.A.P., Benhadi-Marín, J., Mexia, A., Bento, A., Pereira, J.A. 2016. Life-history parameters of *Chrysoperla carnea* s.l. fed on spontaneous plant species and insect honeydews: importance for conservation biological control. BioControl. doi:10.1007/s10526-016-9735-2.

Dinis, A.M., Pereira, J.A., Benhadi-Marín, J., Santos, S.A.P. 2016. Feeding preferences and functional responses of *Calathus Granatensis* and Pterostichus globosus (Coleoptera: Carabidae) on pupae of *Bactrocera oleae* (Diptera: Tephritidae). Bulletin of Entomological Research. doi:10.1017/S0007485316000213

Benhadi-Marín, J., Pereira, J.A., Santos, S.A.P. 2016. Effects of kaolin particle films on the life span of an orb-weaver spider. Chemosphere, 144: 918–924.

Dinis, A.M., Pereira, J.A., Pimenta, M.C., Oliveira, J. Benhadi-Marín, J., Santos, S.A.P. 2015. Suppression of *Bactrocera oleae* (Diptera: Tephritidae) pupae by soil arthropods in the olive grove. Journal of Applied Entomology. doi: 10.1111/jen.12291

José Paulo Sousa

Centre for Functional Ecology, Department of Life Sciences, University of Coimbra, Coimbra, Portugal

Education: PhD in Biology

Research and Professional Experience: On soil ecology, his activities are focused on the development and validation of ecological and biodiversity indicators (using soil invertebrates and trait based approaches) to assess changes in soil quality derived from changes in land-use and landscape fragmentation in agricultural and forest systems. More recently he is interested in evaluating effects of extreme climatic events on soil organisms and service provision in agro-forest systems (in Mediterranean and sub-tropical systems). Regarding soil ecotoxicology, he has been working mainly in the Mediterranean region and the tropics (mainly Brazil) focusing on the development and adaptation of bioassays with soil organisms to assess effects of chemical substances and wastes on soil quality and to integrate in ERA schemes of contaminated soils. He is also involved in evaluating interaction between climate change and pesticides on soil communities and service provisioning

Professional Appointments: Since 2009 is an external expert of the EFSA (European Food Safety Authority) attached to the Scientific Panel for Pesticides, being involved in the elaboration of the new technical guidance documents for pesticide registration in Europe. Since 2014 is has been appointed head of the thematic panel for Pesticides and GMOs for the Portuguese Food Safety Agency (ASAE).

At the University he is the coordinator of the PhD Programme in Biosciences (since 2015) and is the local coordinator of the International Master on Applied Ecology funded by the EU ERASMUS programme (since 2009).

Publications Last Three Years (2014-2016):

Papers in indexed journals

1. Chelinho, S.; Domene, X.; Campana; P.; Andrés, P.; Römbke, J. & Sousa, J.P. (2014) Toxicity of Phenmedipham and Carbendazim to *Enchytraeus crypticus* and *Eisenia andrei* (Oligochaeta) in Mediterranean soils. *Journal of Soils and Sediments*. 14: 584-599.
2. Cesar, R.; Natal-da-Luz, T.; Sousa, J.P.; Colonese, J.; Bidone, E.; Castilhos, Z.; Egler, S. & Polivanov, H. (2014) Disposal of dredged sediments in tropical soils: ecotoxicological effects on earthworms. *Environmental Monitoring and Assessment*. 186: 1487-1497.
3. Creamer, R.E.; Schulte, R.P.O.; Stone, D.; Gal, A.; Krogh, P.H.; Lo Papa, G.; Murray, P.J.; Pérès, G.; Foerster, B.; Rutgers, M.; Sousa, J.P. & Winding, A. (2014) Measuring basal soil respiration across Europe: do incubation temperature and incubation period matter? *Ecological Indicators*.36: 409-418
4. Leitão, S.; Cerejeira, M.J.; Van den Brink, P.J. & Sousa, J.P. (2014) Effects of azoxystrobin, chlorothalonil, and ethoprophos on the reproduction of three terrestrial invertebrates using a natural Mediterranean soil. *Applied Soil Ecology*. 76:124-131.
5. Leitão, S.; Moreira-Santos, M.; Van den Brink, P.J.; Ribeiro, R.; Cerejeira, M.J & Sousa, J.P. (2014) Ethoprophos fate on soil–water interface and effects on non-target terrestrial and aquatic biota under Mediterranean crop-based scenarios. *Ecotoxicology and Environmental Safety*. 103:36-44.
6. Alves, P.R.L.; Cardoso, E.J.B.N.; Martines, A.M.; Sousa, J.P. & Pasini, A. (2014) Seed dressing pesticides on springtails in two ecotoxicological laboratory tests. *Ecotoxicology and Environmental Safety*. 105:65-71.
7. Chelinho, S.; Domene, X.; Andrés, P.; Natal-da-Luz, T.; Norte, C.; Rufino, C.; Lopes, I.; Cachada, A.; Espíndola, E.; Ribeiro, R.; Duarte, A.C. & Sousa, J.P. (2014) Soil microarthropod community testing: A new approach to increase the ecological relevance of effect data for pesticide risk assessment. *Applied Soil Ecology*. 83:200-209.
8. Araújo, C.; Moreira-Santos, M; Sousa, J.P.; Ochoa-Herrera, V.; Encalada, A.C. & Ribeiro, R. (2014). Active avoidance from a crude oil soluble fraction by an Andean Paramo copepod. *Ecotoxicology*. 23:1254-1260

9. Araújo, C.; Moreira-Santos, M; Sousa, J.P.; Ochoa-Herrera, V.; Encalada, A.C. & Ribeiro, R. (2014). Contaminants as habitat disruptors: PAH-driven drift by Andean Paramo stream insects. *Ecotoxicology and Environmental Safety*. 108:89-94
10. Ng, E.-L.; Bandow, C.; Proença, D.N.; Santos, S.; Guilherme, R.; Morais, P.V.; Römbke, J. & Sousa J.P. (2014) Does altered rainfall regime change pesticide effects in soil? A TME study from Mediterranean Portugal on the effects of pyrimethanil to soil microbial communities under extremes in rainfall. *Applied Soil Ecology*. 84:245-253.
11. Cesar, R.; Natal-da-Luz, T.; Rodrigues, F.; Bidone, E.; Castilhos, Z. & Sousa, J.P. (2015). Ecotoxicological assessment of a dredged sediment using bioassays with three species of soil invertebrates. *Ecotoxicology*. 24: 414-423.
12. Cesar R., Natal-da-Luz T., Bidone E., Castilhos Z., Polivanov H. & Sousa J.P. (2015) Disposal of dredged sediments in tropical soils: ecotoxicological evaluation based on bioassays with springtails and enchytraeids. *Environmental Science and Pollution Research*. 22:2916-2924.
13. Gil, F.N; Moreira-Santos, M.; Chelinho, S.; Pereira, C.; Feliciano, J.R.; Leitão, J.H.; Sousa, J.P.; Ribeiro, R. & Viegas, C.A. (2015) Suitability of a *Saccharomyces cerevisiae*-based assay to assess the toxicity of pyrimethanil sprayed soils via surface runoff: Comparison with standard aquatic and soil toxicity assays. *Science of Total Environment*.505: 161-171
14. Zortéa, T.; Baretta, D.; Maccari, A.P.; Segat, J.C.; Boiago, E.S.; Sousa, J.P. & Da Silva, A.S. (2015) Influence of cypermethrin on avoidance behavior, survival and reproduction of *Folsomia candida* in soil. *Chemosphere*. 122: 94-98.
15. Coelho, C.; Branco, R.; Natal-da-Luz, T.; Sousa, J.P. & Morais, P.V. (2015). Evaluation of bacterial biosensors to determine the chromate bioavailability and assess ecotoxicity of soils. *Chemosphere*. 128: 62-69.
16. Rossetti, I.; Bagella, S.; Cappai, C.; Caria, M.C.; Lai, R.; Roggero, P.P.; Martins da Silva, P., Sousa J.P.; Querner, P. & Seddaiu, G. (2015) Influence of isolated cork oak trees on soil features, plant and collembolan assemblages in a Mediterranean wooded grassland. *Agriculture Ecosystems and Environment*.202: 203-216

17. Pereira, C.S., Lopes, I., Sousa, J.P. & Chelinho, S. (2015) Effects of NaCl and seawater induced salinity on survival and reproduction of three soil invertebrates. *Chemosphere.* 135:116-122.
18. Alves, P.R.L.; Natal da Luz, T.; Sousa, J.P. & Cardoso, E.J.B.N. (2015) Ecotoxicological characterization of sugarcane vinasses when applied to tropical soils. *Science of the Total Environment.* 526:222-232.
19. Martins da Silva, P.; Berg, M.; Alves da Silva, A.; Dias, S.; Leitão, P.J.; Chamberlain, D.; Niemelä, J.; Serrano, A.R.M. & Sousa, J.P. (2015) Soil fauna through the landscape window: factors shaping surface- and soil-dwelling communities across spatial scales in cork-oak mosaics. *Landscape Ecology.* 30: 1511-1526.
20. Madani, S.; Coors, A.; Haddioui, A.; Ksibi, M.; Pereira, R.; Sousa, J.P. & Römbke, J. (2015) Effects of contaminated soils from a former iron mine (Ait Amar, Morocco) on enchytraeids (*Enchytraeus bigeminus*) and predatory mites (*Hypoaspis aculeifer*) in standard laboratory tests. *Ecotoxicology and Environmenatl Safety.* 119: 90-97
21. Topping, C.J.; Craig, P.S.; de Jong, F.; Klein, M.; Laskowski, R.; Manachini, B.; Pieper, S.; Smith, R.; Sousa, J.P.; Streissl, F.; Swarowsky, K.; Tiktak, A. & van der Linden, T. (2015). Towards a landscape scale management of pesticides: ERA using changes in modelled occupancy and abundance to assess long-term population impacts of pesticides. *Science of Total Environment.* 537: 159-169
22. Niemeyer, J.C.; Moreira-Santos, M.; Ribeiro, R.; Rutgers, M.; Nogueira, M.A.; Silva, E.M. & Sousa, J.P. (2015) Ecological Risk Assessment of a Metal-Contaminated Area in the Tropics. Tier II: Detailed Assessment. *PLoS ONE* 10(11): e0141772.
23. Stone, D; Blomkvist, P.; Bohse Hendriksen, N.; Bonkowski, M.; Bracht Jørgensen, H.; Carvalho, F.; Dunbar, M.B.; Gardi, C.; Geisen, S.; Griffiths, B.; Hug, A.S.; Jensen, J.; Laudon, H.; Mendes, S.; Morais, P.V.; Orgiazzi, A.; Plassart, P.; Römbke, J.; Rutgers, M.; Schmelz, R.M.; Sousa, J.P.; Steenbergen, S.; Suhadolc, M.; Winding, A.; Zupan, M.; Lemanceau, P. & Creamer, R.E. (2016) A method of establishing a transect for biodiversity and ecosystem function monitoring across Europe. *Applied Soil Ecology.* 97: 3-11
24. Dirilgen, T.; Arroyo, J.; Dimmers, W.J.; Faber, J.; Stone, D.; Martins da Silva, P.; Carvalho, F.; Schmelz, R.; Griffiths, B.; Francisco, R.; Creamer, R.E; Sousa, J.P. & Bolger, T. (2016) Mite community composition across a European transect and its relationships to

variation in other components of soil biodiversity. *Applied Soil Ecology*. 97: 86-97

25. Martins da Silva, P.; Carvalho, C.; Dirilgen, T.; Stone, D.; Creamer, R.E; Bolger, T. & Sousa, J.P. (2016) Traits of collembolan life-form indicate land use types and soil properties across an European transect. *Applied Soil Ecology*. 97: 69-77.
26. Rutgers, M.; Orgiazzi, A.; Gardi, C.; Römbke, J.; Jänsch, S.; Keith, A.; Neilson, R.; Boag, B.; Schmidt, O.; Murchie, A.K.; Blackshaw, R.P.; Pérès, G.; Cluzeau, D.; Guernion, M.; Briones, M.J.I.; Rodeiro, J.; Piñeiro, R.; Díaz Cosín, D.J.; Sousa, J.P.; Suhadolc, M.; Kos, I.; Krogh, P-H.; Faber, J.; Mulder, C.; Bogte, J.J.; Van Wijnen, H.J.; Schouten, A.J. & De Zwart, D. (2016) Mapping earthworm communities in Europe. *Applied Soil Ecology*. 97: 98-111.
27. Creamer, R.E.; Hannula, S.E.; Van Leeuwen, J.P.; Stone, D. Rutgers, M.; Schmelz, R.M.; de Ruiter, P.C.; Hendriksen, N.B.; Bolger, T.; Bouffaud, M.L.; Buee, M.; Carvalho, F.; Costa, D.; Dirilgen, T.; Francisco, R.; Griffiths, B.S.; Griffiths, R.; Martin, F.; Martins da Silva, P.; Mendes, S.; Morais, P.V.; Pereira, C.; Philippot, L.; Plassart, P.; Redecker, D.; Römbke, J.; Sousa, J.P.; Wouterse, M. & Lemanceau, P. (2016) Ecological network analysis reveals the interconnection between soil biodiversity and ecosystem function as affected by land use across Europe. *Applied Soil Ecology*. 97: 112-124.
28. Francisco, R.; Stone, D.; Craemer, R.; Sousa, J.P. & Morais, P.V. (2016) European scale analysis of phospholipid fatty acid composition of soils to establish operating ranges. *Applied Soil Ecology*. 97: 49-60.
29. Reis, F.; Carvalho, F.; Martins da Silva, P.; Mendes, S.; Santos, S.A.P. & Sousa, J.P. (2016) The use of a functional approach as surrogate of Collembola species richness in European perennial crops and forests. *Ecological Indicators*. 61: 676–682.
30. Bandow, C.; Ng, E.L.; Schmelz, R.M.; Sousa, J.P. & Römbke, J. (2016) A TME study with the fungicide pyrimethanil combined with different moisture regimes - Effects on enchytraeids. *Ecotoxicology*. 25: 213-24.
31. Rieff, G.G.; Natal-da-Luz, T.; Sousa, J.P.; Wallau, M.O.; Hahn, L. & Saccol de Sá, E.L. (2016) Collembolans and Mites Communities as a Tool for Assessing Soil Quality: Effect of *Eucalyptus* Plantations on Soil Mesofauna Biodiversity. *Current Science*. 110: 713-719.

32. Niva, C.C.; Niemeyer, J.C.; Silva Júnior, F.M.R.; Nunes, M.E.T.; Sousa, D.L.; Aragão, C.W.S; Sautter, K.D.; Espindola, E.G.; Sousa, J.P. & Römbke, J. (in press) Soil Ecotoxicology in Brazil is taking its course. *Environmental Science and Pollution Research*. DOI 10.1007/s11356-016-6597-1
33. Maccari, A.P; Baretta, D.; Paiano, D.; Leston, S.; Freitas, A.; Ramos, F.; Sousa, J.P. & Klauberg-Filho, O. (in press) Ecotoxicological effects of pig manure on *Folsomia candida* in subtropical Brazilian soils. *Journal of Hazardous Materials*. 314: 113-120.
34. Griffiths, B.S.; Römbke, J.; Schmelz, R.; Scheffczyk, A.; Faber, J.; Bloem, J.; Pérès, G.; Cluzeau, D.; Chabbi, A.; Suhadolc, M.; Sousa, J.P.; Martins da Silva, P.; Carvalho, F.; Mendes, S.; Morais, P; Francisco, R.; Pereira, C.; Bonkowski, M.; Geisen, S.; Bardgett, R.; de Vries, F.; Bolger, T.; Dirilgen, T.; Schmidt, O.; Winding, A.; Hendriksen, N.; Johansen, A.; Philippot, L.; Plassart, P; Bru, D.; Thompson, B.; Griffiths, R.; Keith, A.; Rutgers, M.; Mulder, C.; Hannula, E.; Creamer, R.; Stone, D. (2016) Selecting cost effective and policy-relevant biological indicators for European monitoring of soil biodiversity and ecosystem function. *Ecological Indicators*. 69: 213-223.

Book chapters

1. Viegas, C.A.; Silva, V.; Mateus, C.; Chelinho, S.; Moreira-Santos, M.; Gonçalves, J.; Varela, V.; Ribeiro, R.; Fialho, A.M. & Sousa, J.P. (2014) Bioremediation strategies with *Pseudomonas* sp. ADP in worst-case scenarios of soil contamination with chlorinated s-triazines. In: *Industrial, medical and environmental applications of microorganisms. Current status and trends*. A. Méndez-Vilas (Ed). Wageningen Academic Publishers. 59-64.
2. Brown, G.G.; Niva, C.C.; Zagatto, M.R.G.; Almeida Ferreira, S.; Nadolny, H.S.; Cardoso, G.B.X.; Santos, A.; Martinez, G.A.; Pasini, A.; Bartz, M.L.C.; Sautter, K.D.; Thomazini, M.J.; Baretta, D.; Silva, E.; Antoniolli, Z.I.; Decaëns, T.; Lavelle, P.M.; Sousa, J.P. & Carvalho, F. (2015) Biodiversidade da fauna do solo e sua contribuição para os serviços ambientais. In: *Serviços ambientais em sistemas agrícolas e florestais do Bioma Mata Atlântica*. L.M. Parron, J.R. Garcia, E.B de Oliveira, G.G. Brown e R.B. Prado (Eds). Embrapa. Brasília DF. 122-154.

3. Mendes, S.; Azul, A.M.; Castro, P.; Röembke, J. & Sousa, J.P. (2016) Protecting Soil Biodiversity and Soil Functions: Current Status and Future Challenges. In: *Biodiversity and Education for Sustainable Development*. P. Castro, U.M. Azeiteiro, P. Bacelar Nicolau, W. Leal Filho, M. Azul, A.M. (Eds.). Springer, World Sustainability Series. 249-263. DOI: 10.1007/978-3-319-32318-3_16.

Albino António Bento

Mountain Research Center (CIMO), School of Agriculture,
Polytechnic Institute of Bragança, Portugal.

Education: PhD in Agricultural Sciences and Habilitation

Research and Professional Experience: His main interests of research are focused in the dynamic of pests in Mediterranean woody plants, namely olive, chestnut, almond trees and vineyards in sustainable agriculture. He develops research in the implementation of integrated pest management programmes in those crops.

Professional Appointments: Full Professor in Agricultural Sciences

Publications Last Three Years:

1. Villa, M., Santos, S.A.P., Benhadi-Marín, J., Mexia, A., Bento, A., Pereira, J.A. In press. Life-history parameters of *Chrysoperla carnea* s.l. fed on spontaneous plant species and insect honeydews: importance for conservation biological control. BioControl. http://dx.doi.org/10.1007/s10526-016-9735-2
2. Carocho, M., Barreira, J.C.M., Bento, A., Fernández-Ruíz, V., Morales, P., Ferreira, I.C.F.R. (2016). Chestnut and lemon balm based ingredients as natural preserving agents of the nutritional profile in matured "Serra da Estrela" Food Chemistry 204, Pages 185–193
3. Villa, M., Santos, S.A.P., Mexia, A., Bento, A., Pereira, J.A. 2016. Ground cover management affects parasitism of *Prays oleae* (Bernard). Biological Control. 96: 72-77. http://dx.doi.org/ 10.1016/j.biocontrol.2016.01.012

4. Villa, M.; Marrão, R.; Mexia, A.; Bento, A.; Pereira, J.A., 2016. Are wild flowers and insects honeydews potential food resources for adults of the olive moth, *Prays oleae*? Journal of Pest Science. DOI: 10.1007/s10340-016-0745-8
5. Martins, F.; Pereira, J.A.; Bota, P.; Bento, A.; Baptista, P., 2016. Fungal endophytes communities in above- and belowground olive tree organs and the effect of season and geographic location in their structure. Fungal Ecology, 20, 193-201
6. Sousa, A.; Casal, S.; Malheiro, R.; Lamas, H.; Bento, A.; Pereira, J.A., 2015. Aromatized olive oils: influence of flavouring in quality, composition, stability, antioxidants, and antiradical potential. LWT-Food Science and Technology, 60, 22-28.
7. Oliveira, I.; Pereira, J.A.; Lino-Neto, T.; Bento, A.; Baptista, P., 2015. Plant-mediated effects on entomopathogenic fungi: how the olive tree influences fungal enemies of *Prays oleae*. BioControl, 60, 93-102. Doi: 10.1007/s10526-014-9610-y.
8. Sousa, A.; Malheiro, R.; Casal, S.; Bento, A.; Pereira, J.A., 2015. Optimal harvesting period for cvs. Verdeal Transmontana and Madural, based on antioxidant potential and phenolic composition of olives. LWT - Food Science and Technology, 62, 1120-1126. Doi: doi:10.1016/j.lwt.2015.01.046
9. Sousa, A.; Malheiro, R.; Casal, S.; Bento, A.; Pereira, J.A., 2014. Changes in antioxidant activity and phenolic composition of Cv. Cobrançosa olives through the maturation process. Jornal of Funtional Foods. 11, 20-29.
10. Malheiro, R.; Mendes, P.; Fernandes, F.; Bento, A.; Pereira, J.A., 2014. Natural fermented table olives as source of phenolics, antioxidants and antimicrobials. Foods and Function, 5, 3132-3142.

José Alberto Pereira

School of Agriculture, Polytechnic Institute of Bragança, Portugal.

Education: PhD in Agricultural Sciences

Research and Professional Experience: Dynamic of pests in Mediterranean crops, mainly olive trees and also almond, chestnut, vineyards, and on the influence of agricultural practices in the composition and quality of their products. Nowadays, he is focused on studying strategies and habitat manipulation programs such as the study of alternative food sources and the impact of sustainable products in pests and beneficial entomofauna.

Professional Appointments: Coordinator Professor with Habilitation

Publications Last Three Years:

1. Villa, M., Santos, S.A.P., Benhadi-Marín, J., Mexia, A., Bento, A., Pereira, J.A. In press. Life-history parameters of *Chrysoperla carnea* s.l. fed on spontaneous plant species and insect honeydews: importance for conservation biological control. *BioControl.* http://dx.doi.org/10.1007/s10526-016-9735-2
2. Dinis, A.M., Pereira, J.A., Benhadi-Marín, J., Santos, S.A.P. In press. Feeding preferences and functional responses of *Calathus granatensis* and *Pterostichus globosus* (Coleoptera: Carabidae) on pupae of *Bactrocera oleae* (Diptera: Tephritidae). *Bulletin of Entomological Research.* http://dx.doi.org/10.1017/S0007485316000213
3. Dinis, A.M., Pereira, J.A., Pimenta, M.C., Oliveira, J., Benhadi-Marín, J., Santos, S.A.P. in press. Suppression of *Bactrocera oleae* (Diptera: Tephritidae) pupae by soil arthropods in the olive grove. Journal of Applied Entomology. http://dx.doi.org/10.1111/jen.12291
4. Santos, S.A.P., Mota, L., Malheiro, R., Silva, F., Campos, M., Pinho, P.G., Pereira, J.A. 2016. Changes in volatile compounds of *Dittrichia viscosa* caused by the attack of the gall-forming dipteran *Myopites stylatus. Industrial Crops and Products.* 87: 71–77. http://dx.doi.org/10.1016/j.indcrop.2016.04.002
5. Malheiro, R., Casal, S., Cunha, S.C., Baptista, P., Pereira, J.A. 2016. Identification of leaf volatiles from olive (*Olea europaea*) and their possible role in the ovipositional preferences of olive fly, *Bactrocera oleae* (Rossi) (Diptera: Tephritidae). *Phytochemistry*, 121: 11-19.
6. Benhadi-Marín, J., Pereira, J.A., Santos, S.A.P. 2016. Effects of kaollin particle films on the life span of an orb-weaver spider. Chemosphere. 144: 918-924. http://dx.doi.org/10.1016/j.chemosphere.2015.09.069
7. Villa, M.; Marrão, R.; Mexia, A.; Bento, A.; Pereira, J.A., 2016. Are wild flowers and insects honeydews potential food resources for adults of the olive moth, *Prays oleae*? *Journal of Pest Science.* DOI: 10.1007/s10340-016-0745-8 IF 2.644

8. Martins, F.; Pereira, J.A.; Bota, P.; Bento, A.; Baptista, P., 2016. Fungal endophytes communities in above- and belowground olive tree organs and the effect of season and geographic location in their structure. *Fungal Ecology*, 20, 193-201
9. Rejili, M.; Fernandes, T.; Dinis, A.M.; Pereira, J.A.; Baptista, P.; Santos, S.A.P.; Lino-Neto, T., 2016. A PCR-based diagnostic assay for detecting DNA of the olive fruit fly, *Bactrocera oleae*, in the gut of soil living arthropods. *Bulletin of Entomological Research.*
10. Veloso, A.C.A.; Dias, L.G.; Rodrigues, N.; Pereira, J.A.; Peres, A.M., 2016. Sensory intensity assessment of olive oils using an electronic tongue. *Talanta*, 146, 585-593. 10.1016/j.talanta.2015.08.071.
11. Dias, L.G.; Rodrigues, N.; Veloso, A.C.A.; Pereira, J.A.; Peres, A.M., 2015. Monovarietal extra-virgin olive oil classification: a fusion of human sensory attributes and an electronic tongue. *European Food Research and Technology*. 10.1007/s00217-015-2537-4.
12. Delgado, T.; Santos, L.; Pereira, J.A.; Casal, S.; Ramalhosa, E.; 2015. Effect of drying on colour, proximate composition and drying kinetics of sliced chestnut. *Journal of Food Process Engineering.*
13. Delgado, T.; Pereira, J.A., Ramalhosa, E.; Casal, S., 2015. Effect of hot air convective drying on the fatty acid and vitamin E composition of chestnut (*Castanea sativa* Mill.) slices. *European Food Research and Technology.*
14. Villa, M.; Santos, S.; Mexia, A.; Bento, A.; Pereira, J.A., 2016. Effect of ground cover management on *Prays oleae* (Bernard) and its parasitoids. *Biological Control*. 96, 72-77.
15. Pereira, E.; Ramalhosa, E.; Borges, A.; Pereira, J.A.; Baptista, P., 2015. Dynamics of microbial population during the natural fermentation process of table olives (Negrinha de Freixo Cv.). *Food Microbiology*, 46, 582-586.
16. Sousa, A.; Casal, S.; Malheiro, R.; Lamas, H.; Bento, A.; Pereira, J.A., 2015. Aromatized olive oils: influence of flavouring in quality, composition, stability, antioxidants, and antiradical potential. *LWT-Food Science and Technology*, 60, 22-28.
17. Limón, P.; Malheiro, R.; Acién-Fernández, R.; Fernández-Sevilla, J.M.; Casal, S.; Rodrigues, N.; Cruz, R.; Bermejo, R.; Pereira, J.A., 2015. Virgin olive oils enriched with new antioxidants from the microalgae *Scenedesmus almeriensis:* physicochemical characterization. *Food Chemistry,* 175, 203-211.

18. Fernandes, L.; Pereira, J.A.; Lopéz-Cortés, I.; Salazar, D.; Ramalhosa, E.; Casal, S., 2015. Fatty acid, vitamin E and sterols composition of seed oils from nine different pomegranate (*Punica granatum* L.) cultivars grown in Spain. *Journal of Food Composition and Analysis,* 39, 13-22.
19. Oliveira, I.; Pereira, J.A.; Lino-Neto, T.; Bento, A.; Baptista, P., 2015. Plant-mediated effects on entomopathogenic fungi: how the olive tree influences fungal enemies of *Prays oleae. BioControl*, 60, 93-102. Doi: 10.1007/s10526-014-9610-y.
20. Gama, E.V.S.; Silva, F.; Santos, I.; Malheiro, R.; Fermino Soares, A.C.; Pereira, J.A.; Armond, C., 2015. Homeopathic drugs to control red rot disease in sisal plants. *Agronomy for Sustainable Development,* 35, 646-656. Doi: 10.1007/s13593-014-0255-0.
21. Borges, T.H.; Malheiro, R.; Souza, A.M.; Casal, S.; Pereira, J.A., 2015. Microwave heating induced changes in baru (*Dipteryx alata* Vog.) and soybean crude oils physicochemical properties. *European Journal of Lipid Science and Technology*, 117, 503-513. DOI: 10.1002/ejlt.201400351.
22. Jesus, F.N.; Damasceno, J.C.A.; Barbosa, D.H.S.G.; Malheiro, R.; Pereira, J.A.; Soares, A.C.F., 2015. Control of the banana burrowing nematode using sisal extracts. *Agronomy for Sustainable Development.* 35, 783-791. DOI: 10.1007/s13593-014-0264-z.
23. Cruz, R.; Mendes, E.; Torrinha, A.; Morais, S.; Pereira, J.A.; Baptista, P.; Casal, S., 2015. Revalorization of spent coffee residues by a direct agronomic approach. *Food Research International*, 73, 190-196. DOI:10.1016/j.foodres.2014.11.018.
24. Sousa, A.; Malheiro, R.; Casal, S.; Bento, A.; Pereira, J.A., 2015. Optimal harvesting period for cvs. Verdeal Transmontana and Madural, based on antioxidant potential and phenolic composition of olives. *LWT - Food Science and Technology*, 62, 1120-1126. Doi: doi:10.1016/j.lwt.2015.01.046.
25. Mendez-Tovar, I.; Herrero, B.; Pérez-Magariño, S.; Pereira, J.A.; Asensio-S.-Manzanera, M.C., 2015. By-product of *Lavandula latifolia* essential oil distillation as source of antioxidants. *Journal of Food and Drug Analysis*, 23, 225-233.doi:10.1016/j.jfda.2014.07.003.

26. Malheiro, R.; Casal, S.; Cunha, S.; Baptista, P.; Pereira, J.A., 2015. Olive volatiles from Portuguese cultivars Cobrançosa, Madural and Verdeal Transmontana: role in oviposition preference of *Bactrocera oleae* (Rossi) (Diptera: Tephritidae). *PLoS ONE*, 10(5): e0125070. Doi: 10.1371/journal.pone.0125070.
27. Conde, V.R.; Oliveira, P.F.; Nunes, A.R.; Rocha, C.S.; Ramalhosa, E.; Pereira, J.A., Silva, B.M.; Alves, M.G., 2015. The progression from a lower to a higher invasive stage of bladder cancer is associated with severe alterations in glucose and pyruvate metabolism. *Experimental Cell Research*, 335, 91-98. doi:10.1016/j.yexcr.2015.04.007.
28. Malheiro, R.; Casal, S.; Baptista, P.; Pereira, J.A., 2015. A review of *Bactrocera oleae* (Rossi) impact in olive products: from the tree to the table. *Trends in Food Science & Technology*, 44, 226-242. 10.1016/j.tifs.2015.04.009.
29. Fernandes, L.; Pereira, J.A.; Lopéz-Cortés, I.; Salazar, D.; Ramalhosa, E., 2015. Physico-chemical changes and antioxidant activity of juice, skin, pellicle and seed of pomegranate (Cv, “Mollar de Elche”) at different stages of ripening. *Food Technology and Biotechnology*, 53, 397-406.
30. Malheiro, R.; Ortiz, A.; Casal, S.; Baptista, P.; Pereira, J.A., 2015. Electrophysiological response of *Bactrocera oleae* (Rossi) (Diptera: Tephritidae) adults to olive leaves essential oils from different cultivars and olive tree volatiles. *Industrial Crops and Products*, 77, 81-88. doi:10.1016/j.indcrop.2015.08.046.
31. Malheiro, R.; Casal, S.; Baptista, P.; Pereira, J.A., 2015. Physico-chemical characteristics of *Olea europaea* L. olives and leaves and *Bactrocera oleae* (Rossi) (Diptera: Tephritidae) cultivar oviposition preference. *Scientia Horticulturae* 194, 208–214.
32. Fernandes, L.; Pereira, J.A.; Lopéz-Cortés, I.; Salazar, D.; Ramalhosa, E.; Casal, S., 2015. Lipid composition of seed oils of nine pomegranate (*Punica granatum* L.) cultivars from Spain. International *Journal of Food Studies*, 4, 95-103. Doi: 10.7455/ijfs/4.1.2015.a8.
33. Fernandez-Agullo, A.; Freire, M.S.; Antorrena, G.; Pereira, J.A.; Gonzalez-Alvarez, J., 2014. Effect of the extraction technique and operational conditions on the recovery of bioactive compounds from chestnut (*Castanea sativa*) bur and shell. *Separation Science and Technology,* 49, 267-277.

34. Delgado, T.; Pereira, J.A.; Baptista, P.; Casal, S.; Ramalhosa, E., 2014. Shells role on chestnuts drying kinetics and on fruit`s color and volumetric shrinkage. *Food Research International*, 55, 426-435.
35. Cruz, R.; Gomes, T.; Ferreira, A.; Mendes, E.; Baptista, P.; Cunha, S.; Pereira, J.A.; Ramalhosa, E.; Casal, S., 2014. Antioxidant activity and bioactive compounds of lettuce improved by espresso coffee residues. *Food Chemistry*, 145, 95-101.
36. Cruz, R.; Morais, S.; Mendes, E.; Pereira, J.A., Baptista, P.; Casal, S., 2014. Improvement of vegetables elemental quality by expresso coffee residues. *Food Chemistry,* 148, 294-299.
37. Borges, T.H.; Rodrigues, N.; Souza, A.M.; Pereira, J.A., 2014. Effect of different extraction conditions on the antioxidant potential of baru almonds (*Dipteryx alata* Vog.): comparison to common nuts from Brazil. *Journal of Food and Nutrition Research*, 53, 180-188.
38. Fernandes, L.; Rodrigues, N.; Pereira, J.A.; Ramalhosa, R., 2014. Physico-chemical and sensory characteristics of jellies made from seven grapevine (*Vitis vinifera* L.) varieties. *Acta Agriculturae* Slovenica, 103, 37-47.
39. Martins, F.; Pereira, J.A., Baptista, P., 2014. Oxidative stress response of *Beauveria bassiana* induced by Bordeaux mixture and its influence on fungus growth and development. *Pest management science*, 70, 1220-1227. DOI: 10.1002/ps.3675.
40. Delgado, T.; Marinero, P.; Asensio-S.-Manzanera, M.C.; Asensio, C.; Herrero, B.; Pereira, J.A.; Ramalhosa, E., 2014. Antioxidant activity of twenty wild Spanish *Thymus mastichina* L. populations and its relation with their chemical composition. *LWT - Food Science and Technology*. 57, 412-418.
41. Dias, L.D.; Fernandes, A.; Veloso, A.C.A.; Machado, A.A.S.C.; Pereira, J.A.; Peres, A.M., 2014. Single-cultivar extra virgin olive oil classification using a potentiometric electronic tongue. *Food Chemistry*, 160, 231-239.
42. Teles, S.; Pereira, J.A.; Oliveira, L.M.; Malheiro, R.; Machado, S.S., Lucchese, A.M.; Silva, F., 2014. Organic and mineral fertilization influence on biomass and essential oil production, composition and antioxidant activity of *Lippia origanoides* H.B.K. *Industrial Crops and Products*, 59, 169-176.

43. Teles, S.; Pereira, J.A.; Oliveira, L.M.; Malheiro, R.; Lucchese, A.M.; Silva, F., 2014. *Lippia origanoides* H.B.K. essential oil production, composition, and antioxidant activity under organic and mineral fertilization: effect of harvest time. *Industrial Crops and Products, Crops and Products*, 60, 217-225.
44. Sousa, A.; Malheiro, R.; Casal, S.; Bento, A.; Pereira, J.A., 2014. Changes in antioxidant activity and phenolic composition of Cv. Cobrançosa olives through the maturation process. *Journal of Funtional Foods*. 11, 20-29.
45. Malheiro, R.; Mendes, P.; Fernandes, F.; Bento, A.; Pereira, J.A., 2014. Natural fermented table olives as source of phenolics, antioxidants and antimicrobials. *Foods and Function*, 5, 3132-3142.
46. Peres, A.M.; Veloso, A.C.A.; Pereira, J.A.; Dias, L.G., 2014. Electrochemical multi-sensors devise coupled with heuristic or meta-heuristic selection algorithms for single-cultivar olive oil classification. *Procedia Engineering*, 87, 192-195. DOI: 10.1016/ j.proeng.2014.11.616.

Sónia Alexandra Paiva Santos

Mountain Research Center (CIMO), School of Agriculture,
Polytechnic Institute of Bragança, Portugal.

Education: PhD in Biology

Research and Professional Experience: Her main areas of research include development and validation of ecological and biodiversity indicators and development of strategies and habitat manipulation programs to promote functional biodiversity (natural enemies of pests) in agro-ecosystems.

Professional Appointments: Post-Doc Researcher in Agricultural Sciences

Publications Last Three Years:

1. Villa, M., Santos, S.A.P., Benhadi-Marín, J., Mexia, A., Bento, A., Pereira, J.A. In press. Life-history parameters of *Chrysoperla carnea* s.l. fed on spontaneous plant species and insect honeydews: importance for conservation biological control. *BioControl.* http://dx.doi.org/10.1007/s10526-016-9735-2.

2. Dinis, A.M., Pereira, J.A., Benhadi-Marín, J., Santos, S.A.P. In press. Feeding preferences and functional responses of *Calathus granatensis* and *Pterostichus globosus* (Coleoptera: Carabidae) on pupae of *Bactrocera oleae* (Diptera: Tephritidae). *Bulletin of Entomological Research.* http://dx.doi.org/10.1017/S0007485316000213.
3. Dinis, A.M., Pereira, J.A., Pimenta, M.C., Oliveira, J., Benhadi-Marín, J., Santos, S.A.P. in press. Suppression of *Bactrocera oleae* (Diptera: Tephritidae) pupae by soil arthropods in the olive grove. *Journal of Applied Entomology*. http://dx.doi.org/10.1111/jen.12291.
4. Santos, S.A.P., Mota, L., Malheiro, R., Silva, F., Campos, M., Pinho, P.G., Pereira, J.A. 2016. Changes in volatile compounds of *Dittrichia viscosa* caused by the attack of the gall-forming dipteran *Myopites stylatus*. *Industrial Crops and Products*. 87: 71–77. http://dx.doi.org/10.1016/j.indcrop.2016.04.002.
5. Villa, M., Santos, S.A.P., Mexia, A., Bento, A., Pereira, J.A. 2016. Ground cover management affects parasitism of *Prays oleae* (Bernard). Biological Control. 96: 72-77. http://dx.doi.org/10.1016/j.biocontrol.2016.01.012.
6. Reis, F., Carvalho, F., Martins da Silva, P., Mendes, S., Santos, S.A.P., Sousa, J.P. 2016. The use of a functional approach as surrogate of Collembola species richness in European perennial crops and forests. *Ecological Indicators*. 61: 676-682. http://dx.doi.org/10.1016/j.ecolind.2015.10.019.
7. Benhadi-Marín, J., Pereira, J.A., Santos, S.A.P. 2016. Effects of kaollin particle films on the life span of an orb-weaver spider. Chemosphere. 144: 918-924. http://dx.doi.org/10.1016/j.chemosphere.2015.09.069.
8. Pinheiro, L.A., Torres, L., Raimundo, J., Santos, S.A.P. 2015. Effects of pollen, sugars and honeydew on lifespan and nutrient levels of *Episyrphus balteatus*. *BioContro*l. 60(1):47-57. http://dx.doi.org/10.1007/s10526-014-9621-8.

In: Natural Enemies
Editor: Sónia A.P. Santos
ISBN: 978-1-63485-921-9

Chapter 4

Tracking the Movement of Pest Predators in Arable Crop Landscapes: Implications in Conservation Biological Control Strategies

Filipe Madeira* and Xavier Pons
Department of Crop and Forest Sciences,
Agrotecnio Centre, Universitat de Lleida, Lleida, Spain

Abstract

In many agricultural landscapes perennial and annual crops are adjacent and an exchange of natural enemies between crops and surrounding areas may occur. Traditionally, annual crops are described as "receptor" habitats because they are subject to frequent and severe disturbances and need to be colonised by arthropods every year. By contrast, perennial crops and natural areas are described as "donor" habitats because they are systems with relatively little disturbance and potentially more amenable for the stability of arthropod populations. However, some perennial crops such as alfalfa and meadows undergo periodical cuts during the growing season, which disturb the resident arthropods, forcing them to move to alternative habitats. Moreover, these

* Corresponding Author's address: Department of Crop and Forest Sciences, Agrotecnio Centre, Universitat de Lleida, Rovira Roure 191, 25198 Lleida, Spain. Email: fmadeira@pvcf.udl.cat.

perennial crops need to be recolonised by natural enemies after cutting. In this way, the ecological role played by annual and perennial crops during the season may both be underestimated and overvalued. Clarification of the role that crops and margins play as donor or receptor habitats for natural enemies in arable agricultural landscapes can be useful for decision making in order to improve local habitat management and also to increase the ecosystem services of natural enemies. This chapter is a summary of the studies dealing with this subject in arable crop agroecosystems in the Northeast of the Iberian Peninsula. We report the movement of some of the most representative and abundant predators (Hemiptera: Anthocoridae and Nabidae; Coleoptera: Coccinellidae and Carabidae) between alfalfa (perennial) and maize (annual crop) during the growing season. We also describe the methods used (stable isotope analysis and rubidium as a marker) for determining the movement of natural enemies, their characteristics, as well as advantages and disadvantages of each method. In summary, our studies show that the movement of hemipterans, coccinellids and carabids is species specific and can be influenced by crop management. The perennial crop (alfalfa) acts as source (or donor) of predators for the annual crop (maize) when this crop is at its early vegetative stage.

However, maize acts as a receptor habitat when alfalfa undergoes periodical cuts and as a donor habitat for predators when alfalfa regrows. Maize and herbaceous field margins could therefore act as an alternative/complement to alfalfa strip-cutting, especially in areas with small fields, like in many Mediterranean regions. Later on in the season, alfalfa receives most of the predators that dwell in maize. The association of alfalfa and maize could be very valuable for improving pest management through biological control. This may have significance at both farm and landscape scale. A mosaic of fields of both crops are therefore important for the conservation of the most common and abundant predators.

Keywords: dispersal, Anthocoridae, Nabidae, Coccinellidae, Carabidae, maize, alfalfa, integrated pest management

1. INTRODUCTION

In the last decades, agriculture has been intensified at local and regional scales worldwide, increasing the proportion of monocultures (mainly arable crops), field sizes, and the degree of fragmentation of natural and semi-natural habitats, causing severe changes in agricultural landscape (Baessler & Klotz, 2006; Tscharntke et al., 2005).

The simplification of land use and the destruction of natural areas are considered to be important causes of the decline in farmland biodiversity and may affect the exchange of arthropod natural enemies between non-crop habitats and crops in agricultural landscapes (Bianchi et al., 2006). These changes can affect ecosystem services such as biological control, since non-crop habitats can provide refuge and alternative food resources for a broad spectrum of natural enemies (Andow, 1991; Bianchi et al., 2006; Kleijn et al., 2009; Tscharntke et al., 2007). This is especially true for annual crops, because the success of natural pest control relies on the yearly establishment of natural enemies that immigrate from more stable habitats (Denys & Tscharntke, 2002; Landis & Marino, 1999).

Many agricultural landscapes are mainly formed by a mosaic of annual and perennial crops with some margins and remaining natural areas. In arable crop landscapes, natural enemies can provide valuable biological control services, but for this they need to complete their life cycle in the crop or other habitats in the landscape. Traditionally, annual crops are described as "receptor" habitats because they need to be colonised by arthropods every year due to frequent and severe disturbances. By contrast, perennial crops and natural areas are described as "donor" habitats because they are systems with relatively little disturbance and potentially higher arthropod population stability (Wissinger, 1997). However, some perennial crops such as alfalfa and meadows undergo periodical cuttings during the growing season, which cause a disruption for the resident arthropods, forcing them to move to alternative habitats. Moreover, these perennial crops need to be recolonised by natural enemies when they regrow. In this way, in arable landscapes, the ecological role that annual and perennial crops play for natural enemies during season may be both underestimated and overvalued, because annual crops can act as alternative habitat and donor of natural enemies after the cutting of perennial crops, which then become "receptors."

Crop manipulation aiming at enhancing the performance and abundance of the existing natural enemies is one of the key strategies in conservation biological control (Landis et al., 2000), which is the only cost-effective pest control method in the arable crop conditions of many Mediterranean regions (Pons et al., 2013). Detailed information about crop/habitat functionality (as donor *or* receptor of predators) and about the movement patterns of predator populations can be useful for decision making in local habitat management and also to improve conservation biological control strategies by increasing the potential of natural enemies (Barbosa & Wratten, 1998; Dent, 2000; Landis et al., 2000). Therefore, there is the need to study the dynamics of the movement of the most abundant predators present in the arable crop landscape.

This chapter is a summary of the studies dealing with this subject in arable crop agroecosystems in the northeast of the Iberian Peninsula. We report the movement between alfalfa (perennial) and maize (annual crop) of some of the most representative and abundant predators (Hemiptera: Anthocoridae and Nabidae and Coleoptera: Coccinellidae and Carabidae) during the growing season. We also briefly describe the methods used (rubidium as a marker and stable isotope analysis) for determining the movement, their characteristics and some advantages and disadvantages of each method.

2. Techniques for Tracking Insect Movement

In the last decades, a wide variety of techniques for tracking arthropod movement have been developed (Lavandero et al., 2004; Reynolds et al., 1997). Such techniques should be environmentally safe, cost-effective, easy to use and persistent without affecting the behavior of insects (Hagler & Jackson, 2001). However, the selection of a technique should depend on the insect species, environmental conditions, the nature of the experiment and the study objectives (Hagler & Jackson, 2001).

Depending on the type of study, insects can be marked on purpose or they can mark themselves through contact with marking materials that naturally occur in their environment or materials that have been strategically added (Reynolds et al., 1997). The markers can be external, where the mark is usually applied into the insect surface and recognised by visual methods; or internal, where the recognition of markers requires specialised equipment (Lavandero et al., 2004). Although external markers are commonly used in mark-release-recapture studies (Hagler & Jackson, 2001) they can be difficult to apply on small animals (Hobson & Norris, 2008; IAEA, 2009). Nowadays, in mark-recapture studies, the most commonly employed techniques are the analysis of rare earth elements and stable isotope analysis (Scherber et al., 2012).

2.1. Rubidium Marking and Stable Isotope Analysis for Tracking Insect Movement

Rubidium in its chloride form is the most common rare elemental marker used for tracking insects (Hagler & Jackson, 2001). This element replaces the potassium in the tissues of insects feeding on a treated source (Berry et al., 1972). Rubidium is present in nature at very low concentrations and techniques using it have to raise the natural level through an application of an aqueous solution of rubidium chloride (RbCl) (Berry et al., 1972). The solution can be applied in a specific habitat where it is absorbed by soil/plants and moves between trophic levels. Plants may be marked by augmentation of environmental rubidium by spraying RbCl on soil or foliage (Guillebeau et al., 1993). Herbivores can be marked by direct contact when the marker is applied, or later by feeding on or contact with rubidium-marked plants (Fernandes et al., 1997; Graham et al., 1978). Predators and parasitoids can be marked through direct application, feeding on rubidium-marked prey and/or feeding on floral nectar or pollen from rubidium-marked plants and by direct contact with rubidium-marked soil/plants (Graham et al., 1978; Johnson & Reeves, 1995; Lavandero et al., 2004). This technique offers many advantages over other insect marking procedures. Rubidium is not radioactive, so it is safe for workers and the environment and at moderate concentrations it has no adverse effect on the insects' physiology or behavior (Hagler & Jackson 2001). This technique can also be used for marking small predators and parasitoids (e.g., Corbett et al., 1996; Prasifka et al., 2001) and the fact that insects can be marked in the laboratory and then released in the field allows to determine the exact distance that they have moved during a time period. Some disadvantages of this technique include the need of using expensive detection equipment, technical expertise and a specific laboratory protocol (Stimmann, 1991). Arthropods remain marked only for a few days, the marking decays with change of diet and there is the application needs to be repeated from time to time (Wolfenbarger et al., 1982).

Many studies have used this technique to measure the movement of herbivores and natural enemies in the field (e.g., Albajes et al., 2004; Fernandes et al., 1997; Perović et al., 2011; Prasifka et al., 2001; Scarratt et al., 2008; Tillman et al., 2007).

Plants may be used as markers for tracking insects through their natural concentration in stable isotopes of chemical elements. This is especially true when the system is mainly formed by two clearly isotopically distinct crops such as C3 and C4 plants, because they fix CO_2 using different pathways. The carbon stable isotope ratio ($\delta^{13}C$) of C4 plants is distinctly higher than the one of C3 plants. In particular, $\delta^{13}C$ has recently been proven to be suitable for tracing the plant origin of phytophagous (Schallhart et al., 2011; Traugott et al., 2008; Vialatte et al., 2006) and predatory insects (di Lascio et al., 2016; Forbes & Gratton, 2011; Madeira et al., 2014; Ouyang et al., 2012; Prasifka et al., 2004; Prasifka & Heinz, 2004). The $\delta^{13}C$ is transferred with low fractionation (0‰ ± 1‰) between trophic levels (DeNiro & Epstein, 1981;

Ostrom et al., 1997; Post, 2002). The herbivores that feed on C3 or C4 plants reflect the overall isotopic composition of their food, and the predators that feed on these herbivores will subsequently reflect the isotopic concentration of their prey (DeNiro & Epstein, 1978; Oelbermann & Scheu, 2002). The isotopic composition of predators moving between and feeding on isotopically distinct crops such as maize (C4) and alfalfa (C3) will gradually move from the original source towards the signature of the new food source (Madeira et al., 2013; Ostrom et al., 1997; Ouyang et al., 2012; Prasifka et al., 2004). Moreover, nitrogen stable isotope ratios ($\delta^{15}N$), which are generally used as an indicator of the trophic position (DeNiro & Epstein, 1981), can also improve information about the incorporation of prey from different resources, with an enrichment of 3.4‰ ± 1‰ between predators and their prey (Post, 2002).

This technique has the advantage that the plants themselves may be used as markers, avoiding the application of any material/element. The change of food source can be detected within the first five days after the diet change but at the same time traces of the old source remain for 20-25 days after the diet change (Madeira et al., 2013; Ostrom et al., 1997; Ouyang et al., 2012; Prasifka et al., 2004). Some disadvantages of this technique are also the need of expensive detection equipment, technical expertise and a specific laboratory protocol.

3. Movement of Pest Predators between Alfalfa and Maize in a Mediterranean Crop System – Case Studies

In the irrigated land of the northeastern Iberian Peninsula, the arable crop landscape is formed mainly by annual crops such as winter cereals and maize and perennial crops such as alfalfa, building a mosaic with field margins, uncultivated lands, woody areas and fruit orchards. Growth cycles of these arable crops partially overlap: alfalfa and winter cereals in spring, alfalfa and maize in summer, and all three crops in late spring and autumn. Only winter cereals and maize share some herbivores, but similar natural enemies can be observed within all three crops (Pons & Eizaguirre, 2009). During summer, the arable crop landscape is mainly shaped by alfalfa and maize, which occupy a surface of approximately 110.000 and 105.000 ha, respectively (Anuario Estadístico de Catalunya and Aragón, 2014). These two crops share a broad spectrum of predators such as hemipterans (Anthocoridae and Nabidae), coleopterans (Coccinellidae, Carabidae and Staphylinidae), neuropterans (Chrysopidae), dipterans (Syrphidae and Cecidomyiidae) and several spider families (Pons & Eizaguirre, 2009). In arable crops, these predators can keep some pest levels below an economic threshold if their action is not disturbed (Pons & Eizaguirre, 2009).

In this chapter, we provide an overview of the movements between maize and alfalfa of the generalist predators *Orius majusculus* Reuter, *Orius niger* Wolf (Hemiptera: Anthocoridae) and *Nabis provencalis* Remane (Hemiptera: Nabidae) and the specialised aphidophagous beetle species *Coccinella septempunctata* L., *Hippodamia variegata* Goeze and *Propylea quatuordecimpunctata* L. (Coleoptera: Coccinellidae) since they are among the most common and abundant plant-dwelling predators in both maize (Albajes et al., 2003; De la Poza et al., 2005; Pons & Eizaguirre, 2009) and alfalfa (Pons & Eizaguirre, 2009; Pons et al., 2005). On the other hand, we also provide an overview over the movements of carabid beetles (Coleoptera: Carabidae), the most common and abundant ground-dwelling predators in maize (Albajes et al., 2009; De la Poza et al., 2005; Pons and Eizaguirre, 2009) and in alfalfa (Núñez, 2002). We selected *Calathus fuscipes* Goeze, *Poecilus cupreus* DeGeer, *Bembidion lampros* Herbst and *Pseudoophonus rufipes* L., since these four species represented 90% of the total catches of carabids in alfalfa and maize (Madeira & Pons, 2016).

For tracking the movement of plant-dwelling predators we used carbon and nitrogen stable isotopes and rubidium as markers for carabids. These experiments were carried out in fields located in Lleida province (Catalonia, northeastern Iberian Peninsula). The size and proximity of alfalfa and maize fields and the characteristics of the surrounding landscape at the study locations are described in Madeira et al. (2014), Madeira & Pons (2016) and di Lascio et al. (2016). The movement of plant-dwelling predators was assessed four times in the course of the growing season: 1) when maize was at its early vegetative stage, 2) just after alfalfa cutting, 3) when alfalfa regrew after cutting and 4) after maize harvesting or plant drying. The movement of ground-dwelling predators was assessed before and after alfalfa cutting and after maize harvesting.

During the *maize early vegetative stage*, all plant-dwelling predators which colonized maize had moved from alfalfa, which acts as a donor crop at this period of the season. Nevertheless, *O. niger*, *N. provencalis*, *H. variegata* and *P. quatuordecimpunctata* also moved from maize to alfalfa, presenting bidirectional movement between crops. This indicates that species from the same family show a different dispersal pattern (Figure 1).

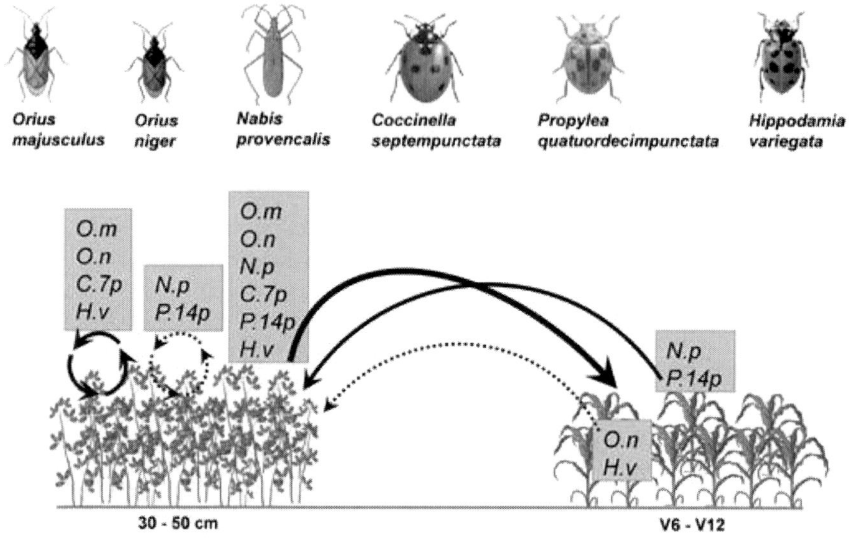

Figure 1. Schematic representation of the movement of the plant-dwelling predators Orius majusculus (O.m), Orius niger (O.n), Nabis provencalis (N.p), Coccinella septempunctata (C.7p), Hippodamia variegata (H.v) and Propylea quatuordecimpunctata (P.14p) between alfalfa and maize during the early maize vegetative stage (maize was at V6-V12 stage according to Ritchie et al., (2005) and alfalfa was between 30-50 cm height). Arrows represent the direction of movement and circle symbols represent establishment in the crop, thickness of the arrows and circles indicates flux intensity.

According to the turnover reported for coccinellids (Ostrom et al., 1997; Ouyang et al., 2012; Prasifka et al., 2004) and heteropterans (Madeira et al., 2013), plant-dwelling predators found in maize have moved from alfalfa 20-25 days before, while most of these predators in alfalfa showed a stable population.

These results in part confirm that alfalfa acts as a reservoir and a source of predators that colonise annual crops (Núñez, 2002; Summers, 1998; Pons et al., 2005). The observed movement pattern of the *Orius* species is in agreement with the hypothesis of Pons et al. (2005), who proposed that they spend autumn and winter in alfalfa, which then becomes a source of individuals dispersing to maize in spring. Moreover, our results suggest that in Mediterranean agricultural systems, where the presence of alfalfa is common, colonisation of other crops by *Orius* spp. occurred earlier than in other Central European agroecosystems (Veres et al., 2012). A dispersal of ladybirds from alfalfa to maize according to their field abundance was also suggested (Pons & Eizaguirre, 2009).

During the season, alfalfa is subject to cuttings that cause temporary but drastic changes to the system. Just after alfalfa cuttings, we observed movement from alfalfa to maize in all plant-dwelling and ground-dwelling predators. However, species showed different movement patterns. In the case of plant-dwelling predators we observed that most of *C. septempunctata*, *O. niger* and *N. provencalis* collected in maize came from alfalfa while only few of *O. majusculus* and *P. quatuordecimpunctata* had moved from alfalfa. This suggests that most of the populations of this two species were scarce in alfalfa and probably already well-established in maize (Figure 2A).

Bidirectional movement after alfalfa cutting was observed in the case of carabids but with higher intensity from alfalfa to maize than vice-versa, except for *B. lampros* which seemed to be more stable in alfalfa (Figure 2B). In other studies, crop management practices (e.g., cutting) appear to trigger the movement of carabids to alternative habitats (Eyre et al., 2013; Ribera et al., 2001; Thorbek & Bilde, 2004). Curiously, after alfalfa cutting, we also observed movements from maize to alfalfa and to the margins between crops (Figure 2B). This may suggest that individuals remained in the margin until alfalfa regrew and started to recolonise alfalfa as soon as it became suitable again.

Globally, these findings show that alfalfa cutting triggers the movement of predators to maize and that this crop acts as a receptor habitat. The margin between crops can also act as a receptor habitat for carabids after alfalfa cutting.

After cutting, alfalfa regrows quickly and three weeks later plants can already reach a height of 30-50 cm. During this period, when alfalfa and maize were growing simultaneously again, *O. niger*, *N. provencalis* and *P. quatuordecimpunctata* and all carabids presented bidirectional movements. All plant-dwelling predators, except *C. septempunctata*, recolonised alfalfa from maize. At this time of the season there were well established populations of heteropterans in maize, and *C. septempunctata*, *H. variegata* and *N. provencalis* in alfalfa. *O. majusculus* and *C. septempunctata* showed inverse movement patterns between crops (Figure 3A).

The directional movement of carabids from alfalfa to maize was biased for *C. fuscipes* and *P. rufipes* and the contrary was observed for the other two species. *C. fuscipes*, *P. rufipes* and *P. cupreus* continued inhabiting the margin when both crops were growing, while *B. lampros* appeared to only use the margin if alfalfa was cut (Figure 3B).

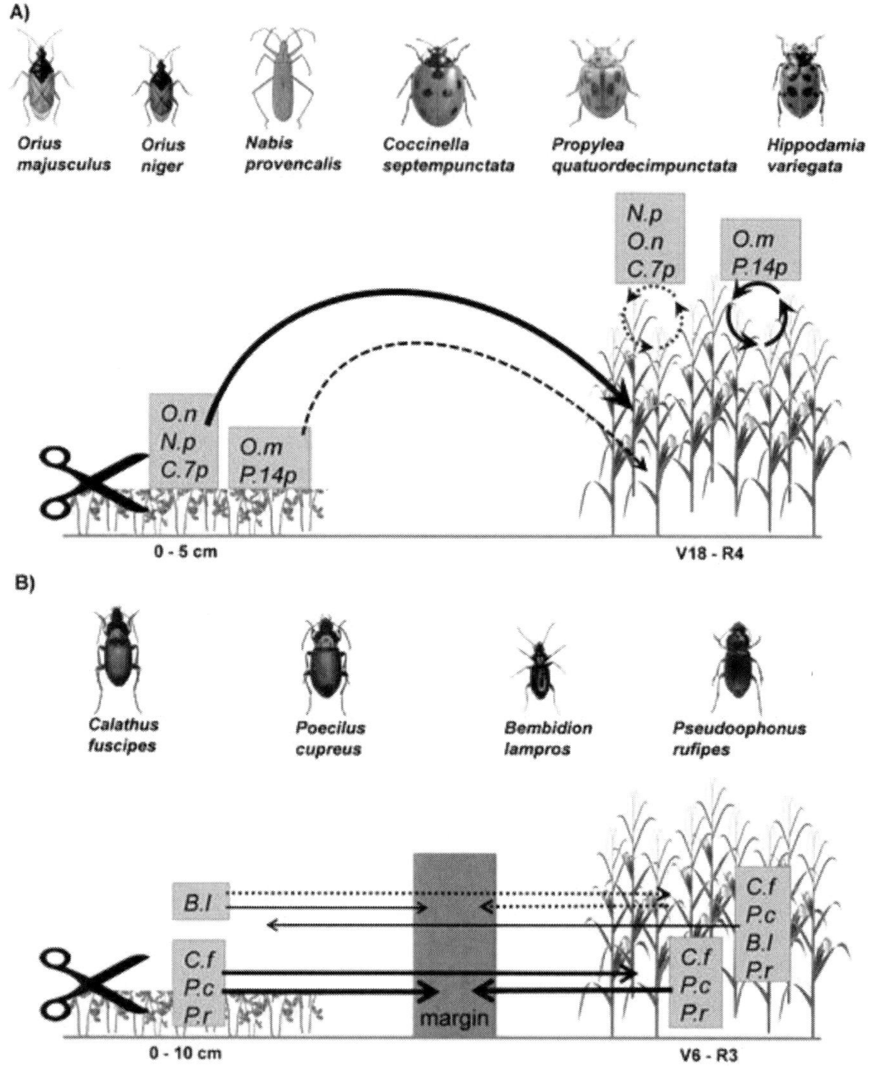

Figure 2. Schematic representation of the movement of the (A) plant-dwelling predators Orius majusculus (O.m), Orius niger (O.n), Nabis provencalis (N.p), Coccinella septempunctata (C.7p), Hippodamia variegata (H.v), Propylea quatuordecimpunctata (P.14p) and the (B) ground-dwelling predators Calathus fuscipes (C.f), Poecilus cupreus (P.c), Bembidion lampros (B.l) and Pseudoophonus rufipes (P.r) between alfalfa and maize after alfalfa cutting (maize was at V6-R4 stage according to Ritchie et al., (2005) and alfalfa was between 0-10 cm height). Arrows represent the direction of movement and circle symbols represent establishment in the crop, thickness of the arrows and circles indicates flux intensity.

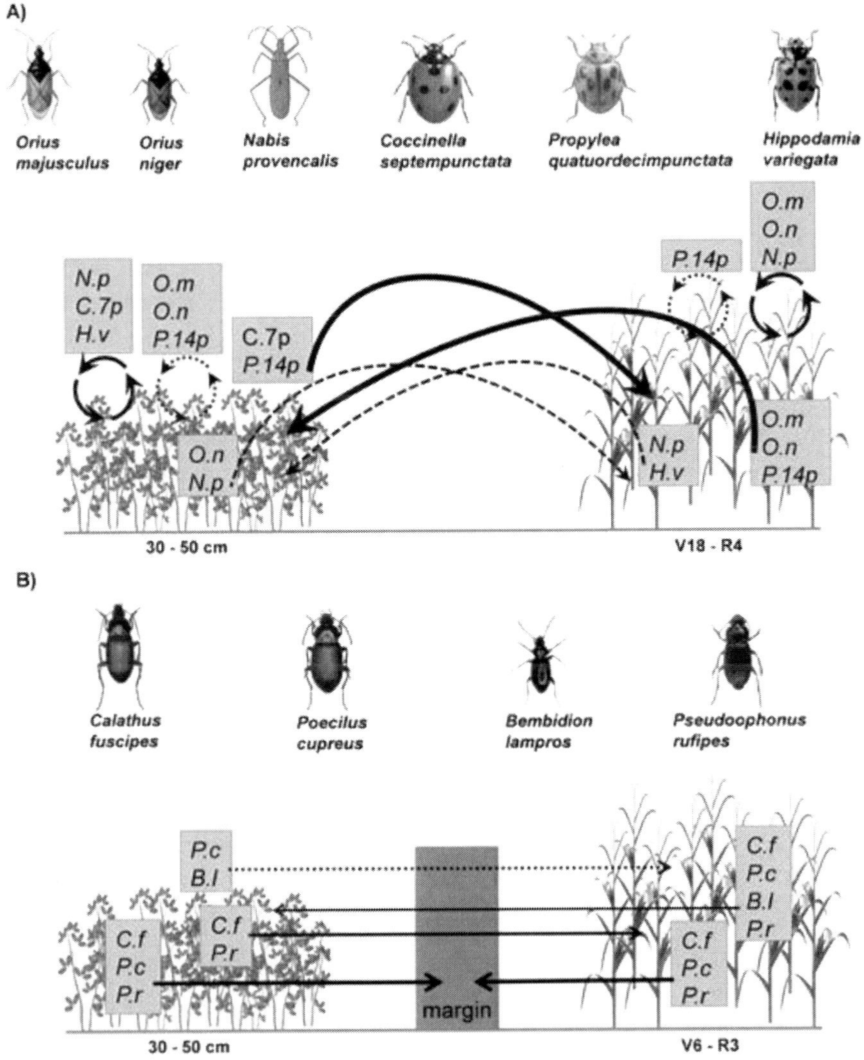

Figure 3. Schematic representation of the movement of the (A) plant-dwelling predators Orius majusculus (O.m), Orius niger (O.n), Nabis provencalis (N.p), Coccinella septempunctata (C.7p), Hippodamia variegata (H.v), Propylea quatuordecimpunctata (P.14p) and the (B) ground-dwelling predators Calathus fuscipes (C.f), Poecilus cupreus (P.c), Bembidion lampros (B.l) and Pseudoophonus rufipes (P.r) between alfalfa and maize when alfalfa regrew after cutting (maize was at V6-R4 stage according to Ritchie et al., 2005 and alfalfa was between 30-50 cm height). Arrows represent the direction of movement and circle symbols represent establishment in the crop, thickness of the arrows and circles indicates flux intensity.

Overall, these findings suggest that maize adjacent to alfalfa enables predators to recolonize alfalfa when plants regrow and points out the role of maize as a source crop.

When maize dried or was harvested, there was movement of predators (except *C. septempunctata* and *P. rufipes*) from maize to alfalfa (Figure 4A and Figure 4B). *C. fuscipes* was the only carabid that moved to the margin after maize harvesting. *O. majusculus* showed a different movement pattern compared to the rest of plant-dwellings predators. This predator was the last to abandon maize because it is the only plant-dwelling species that is not well established in alfalfa at this time of the season (Figure 4A). The return of *Orius* species to alfalfa confirms the hypothesis of Pons et al. (2005) indicated above. At this time of the season, alfalfa acts as receptor for most of predators that were in maize except for *C. septempunctata* which never returns to alfalfa after moving to maize.

4. Effect of Crop Management on the Movement of Predators

The movement and dispersal of predators inhabiting alfalfa and maize are subject to the effect of crop management practices. Because the patterns of movement have been demonstrated (in our studies) to be species-specific, these practices and the timing when they are performed will not equally affect all predators. Moreover, not all predators have the same affinity for both crops. While *C. septempunctata*, *N. provencalis*, *O. niger* and *B. lampros* seem to be more faithful for alfalfa, *O. majusculus* and *C. fuscipes* seem to be more prone to inhabit maize (di Lascio et al., 2016; Madeira & Pons, 2016; Madeira et al., in preparation).

Two of the practices that can affect the movement of predators and that occur in every growing season are alfalfa cutting and maize harvesting. It has been demonstrated that both of them trigger the movement of predators to alternative habitats.

We observed that after alfalfa cutting all plant-dwelling predators and carabids inhabiting alfalfa moved to maize. However, the timing of the cutting can also affect the survival, and consequently the dispersal of some predator species, e.g., *C. septempunctata* (di Lascio et al., 2016). In the study region, at the time of the second cutting, populations of this ladybird are mainly dominated by larvae (Pons, pers. obs.) and many of them are killed by the cutting process. If this second cutting was delayed, population sizes could be increased, since larvae could reach the adult stage. On the contrary, at this time of the season, populations of *P. quatuordecimpunctata* are better established in maize, and cutting will affect it to a lesser extent (di Lascio et al., 2016). On the other hand, a catastrophic management practice such as alfalfa cutting does not seems to have serious consequences on the movement and recolonisation of alfalfa by carabids (Madeira & Pons, 2016) and heteropterans (Pons et al., 2009, Madeira et al., in preparation).

Because alfalfa cutting time is mainly determined by crop development, and delaying the date of cutting could be negative for the yield, alfalfa strip-cutting (Hossain et al., 2002; Landis et al., 2000) could be a strategy for conserving predators. Núñez et al. (2007) reported significantly higher aphid predator densities in alfalfa fields with strip bands than in fields that were cut entirely in the northeastern Iberian Peninsula. *N. provencalis* and *O. niger* can recolonize alfalfa after cutting from own or nearby alfalfa fields. Therefore, strip cutting could contribute to recolonisation (Madeira et al., in preparation). However, in regions where field size is low, this practice is not easy to apply (Pons & Eizaguirre, 2009) and a mosaic landscape that includes alfalfa fields with gradual cutting dates and/or adjacent maize fields could be an alternative or a complement to alfalfa strips. The grassy margins between crops or bordering fields can also act as a complement to alfalfa strips for carabids (Madeira & Pons, 2016).

Similarly to what happens after alfalfa cutting, most predators move to alfalfa when maize dries or is harvested, except *C. septempunctata* which abandon maize in order to find food resources in refuges other than alfalfa (di Lascio et al., 2016).

Insecticide sprayings can affect predators, directly affecting their survival by their toxic action or indirectly through changes in prey availability. This last effect may affect movements. Insecticide treatments applied in maize at sowing or pre-emergence are common in the studied region. They affect the abundance of some sap-sucking herbivores, especially aphids, in the beginning of maize growth (Pons & Albajes, 2002). This makes coccinellid establishment difficult and individuals arriving to maize will probably have to migrate to other plants infested with aphid for feeding due to the low quality of the habitat (Hodek & Michaud, 2008). This seems to affect *C. septempunctata* in particular, probably due to its greater requirements of aphid density compared to the smaller coccinellid species (Dixon, 2007; Honek et al., 2008). Furthermore, the abundances of *Orius* sp. and *Nabis* sp. have been reported to be correlated with the availability of aphids and leafhoppers (Albajes et al.,

2011). Therefore, the movement of heteropterans can also be affected by insecticide spraying at that time.

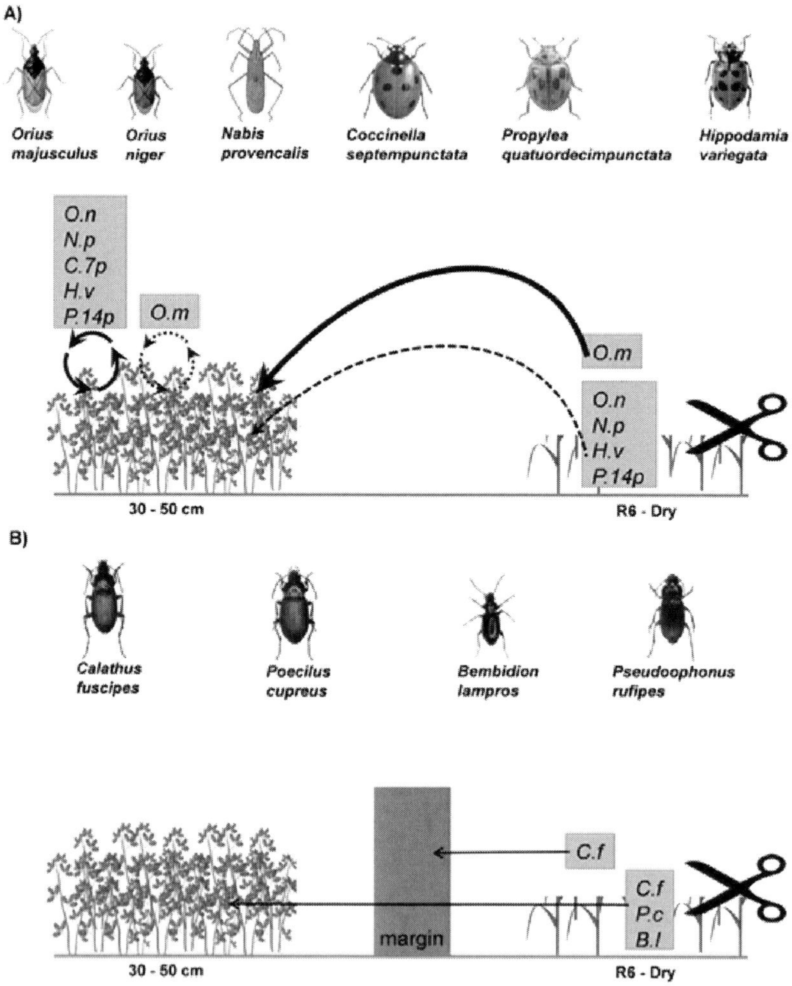

Figure 4. Schematic representation of the movement of the (A) plant-dwelling predators Orius majusculus (O.m), Orius niger (O.n), Nabis provencalis (N.p), Coccinella septempunctata (C.7p), Hippodamia variegata (H.v), Propylea quatuordecimpunctata (P.14p) and (B) ground-dwelling predators Calathus fuscipes (C.f), Poecilus cupreus (P.c), Bembidion lampros (B.l) and Pseudoophonus rufipes (P.r) between alfalfa and maize after maize harvesting or plant drying (maize was at R6-Dry stage according to Ritchie et al., (2005) and alfalfa was between 30-50 cm height). Arrows represent the direction of movement and circle symbols represent establishment in the crop, thickness of the arrows and circles indicates flux intensity.

Aerial insecticide sprayings applied to maize later in the season, mainly against the European corn borer, *Ostrinia nubilalis* Hübner (Lepidoptera: Crambidae), may kill *O. majusculus* (Al-Deeb et al., 2001; Filho et al., 2002; de la Poza et al., 2005) and obstruct the dispersal of this species from maize to alfalfa. On the other hand, in the study area, insecticide sprayings on alfalfa are usually restricted to the alfalfa weevil, *Hypera postica* Gyllenhall (Coleoptera: Curculionidae) and are applied during the growing period before the first cutting. Therefore, the effect of this practice will have very little influence on predators and their movement between alfalfa and maize because at this time of the season only few fields of maize have already been sown.

In the northeast of the Iberian Peninsula, genetically modified maize that contains the toxin of *Bacillus thuringiensis* (Bt maize) is predominant. Bt maize allows to reduce maize corn borer damage and to reduce the negative effects of the insecticide spraying (Fitt, 2008). It has been demonstrated that Bt maize does not negatively affect *Orius* sp. in the field (Comas et al., 2014) and that positive effects on the performance of *O. majusculus* could occur (Lumbierres et al., 2012). Thus, Bt maize does not negatively affect the movement pattern and capacity of *O. majusculus*. The use of Bt maize which requires few or no insecticide applications at all could favour an increase in predator populations as observed in cotton (Lu et al., 2012). These benefits and a hypothetic increase in predator abundances might provide additional biocontrol services due to the movement of predators from maize to neighbouring crops.

Other practices such as herbicide treatments causing moderate changes in weed densities within maize did neither affect the abundance of most predators inhabiting maize (Albajes et al., 2009), nor the dispersal of *O. majusculus* (Madeira & Pons, 2015).

CONCLUSION

The case studies presented here point out that:

1. The movement of predators between alfalfa and maize occurs naturally when both crops are present in the landscape. It is trigged by crop management practices such as alfalfa cutting or maize harvesting.
2. The movement pattern of plant-dwelling and ground-dwelling predators is species-specific.

3. Alfalfa facilitates early colonization of maize by heteropterans and coccinellids, while maize later provides these predators and carabids for recolonising alfalfa after cutting. Maize and herbaceous field margins could therefore act as an alternative/complement to alfalfa strip-cutting, especially in areas with small fields, as in many Mediterranean regions. Later in the season, alfalfa receives most predators that were dwelling in maize.
4. The association of alfalfa and maize could be very valuable for improving pest management through biological control. This may have significance at both farm and landscape scale. Farms with adjacent alfalfa and maize fields and landscapes including a mosaic of fields of both crops are therefore important for the conservation of the most common and abundant predator species.

References

Albajes, R., López, C. & Pons, X. (2003). Predatory fauna in cornfields and response to imidacloprid seed treatment. *Journal of Economic Entomology, 96*, 1805–1813.

Albajes, R., Eras, J., López, C., Ferran, X., Vigatà, J. & Eizaguirre, M. (2004). Testing rubidium marking for measuring adult dispersal of the corn borer *Sesamia nonagrioides*: first results. *IOBC/WPRS Bulletin, 27*, 15–22.

Albajes, R., Lumbierres, B. & Pons, X. (2009). Responsiveness of arthropod herbivores and their natural enemies to modified weed management in corn. *Environmental Entomology, 38*, 944–54.

Albajes, R., Lumbierres, B. & Pons, X. (2011). Two heteropteran predators in relation to weed management in herbicide-tolerant corn. *Biological Control, 59*, 30–36.

Al-Deeb, M. A., Wilde, G. E. & Higgins, R. A. (2001). No effects of *Bacillus thuringiensis* on the predator *Orius insidiosus* (Hemiptera: Anthocoridae). *Environmental Entomology, 30*, 625-629.

Andow, D. (1991). Vegetational diversity and arthropod population response. *Annual Review of Entomology, 36*, 561–586.

Anuario Estadístico de Catalunya. Agricultura. http://www.idescat.cat/pub/?id=aec&n=444. Accessed the April 14, 2016.

Anuario Estadístico Agrario Gobierno de Aragon http://www.aragon.es/DepartamentosOrganismosPublicos/Departamentos/DesarrolloRuralSostenibilidad/AreasTematicas/EstadisticasAgrarias/ci.ANUARIO_ESTADISTICO_AGRARIO.detalleDepartamento?channelSelected=1cfbc8548b73a210VgnVCM100000450a15acRCRD#section2. Accessed the April 14, 2016.

Baessler, C. & Klotz, S. (2006). Effects of changes in agricultural land-use on landscape structure and arable weed vegetation over the last 50 years. *Agriculture, Ecosystems and Environment,* 115, 43–50.

Barbosa, P. & Wratten, S. D. (1998). Influence of plants on invertebrate predators: implications to conservation biological control. In: P, Barbosa (Ed.), *Conservation Biological Control* (pp. 83–100). San Diego: Academic Press.

Berry, W. L., Stimmann, M. W. & Wolf, W. W. (1972). Marking of native phytophagous insects with rubidium: A proposed technique. *Annals of the Entomological Society of America, 27*, 236–238.

Bianchi, F. J. J. A., Booij, C. J. H. & Tscharntke, T. (2006). Sustainable pest regulation in agricultural landscapes: a review on landscape composition, biodiversity and natural pest control. *Proceedings of the Royal Society B: Biological Sciences, 273*, 1715–1727.

Comas, C., Lumbierres, B., Pons, X. & Albajes, R. (2014). No effects of *Bacillus thuringiensis* maize on non target organisms in the field in southern Europe: a meta-analysis of 26 arthropod taxa. *Transgenic Research, 23*, 135–143.

Corbett, A. Murphy, B. C. Rosenheim, J. & Bruins, P. (1996). Labeling and egg parasitoid, *Anagrus epos* (Hymenoptera: Mymaridae), with rubidium within an overwintering refuge. *Environmental Entomology, 25*, 29–38.

De la Poza, M., Pons, X., Farinós, G. P., López, C., Ortego, F., Eizaguirre, M., Castañera, P. & Albajes, R. (2005). Impact of farm-scale Bt maize on abundance of predatory arthropods in Spain. *Crop Protection, 24,* 677–684.

DeNiro, M. J. & Epstein, S. (1978). Influence of diet on the distribution of carbon isotopes in animals. *Geochimica et Cosmochimica Acta, 42,* 495–506.

DeNiro, M. J. & Epstein, S. (1981). Influence of diet on the distribution of nitrogen isotopes in animals. *Geochimica et Cosmochimica Acta, 45*, 341–351.

Dent, D. (2000). *Insect Pest Management.* London: CAB International.

Denys, C. & Tscharntke, T. (2002). Plant-insect communities and predator-prey ratios in field margin strips, adjacent crop fields, and fallows. *Oecologia, 130*, 315–324.

Dixon, A. F. G. (2007). Body size and resource partitioning in ladybirds. *Population Ecology, 49*, 45–50.

di Lascio, A., Madeira, F., Costantini, M. L., Rossi, L. & Pons, X. (2016). Spillover between alfalfa and maize of three ladybird species (Col. Coccinellidae) revealed by stable isotope analysis. *BioControl, 61*, 35–46.

Eyre, M. D., Luff, M. L. & Leifert, C. (2013). Crop, field boundary, productivity and disturbance influences on ground beetles (Coleoptera, Carabidae) in the agroecosystem. *Agriculture, Ecosystems and Environment, 165*, 60–67.

Fernandes, O., Wright, R., Baumgarten, K. & Mayo, Z. (1997). Use of rubidium to label *Lysiphlebus testaceipes* (Hymenoptera: Braconidae), a parasitoid of greenbugs (Homoptera: Aphididae), for dispersal studies. *Environmental Entomology, 26*, 1167–1172.

Filho, M. M., Della Lucia, T. M. C., Cruz, I., Guedes, R. N. C. & Galvão, J.C.C. (2002). Chlorpyrifos spraying of no-tillage corn during tasselling and its effect on damage by *Helicoverpa zea* (Lep., Noctuidae) and on its natural enemies. *Journal of Applied Entomology, 126*, 422-430.

Fitt, G. P. (2008). Have Bt crops led to changes in insecticide use patterns and impacted IPM? In: J. Romeis, A. M. Shelton & G. C. Kennedy (Eds.), *Integration of Insect-Resistant Genetically Modified Crops Within IPM Programs* (pp: 303-328). Berlin: Springer Science + Business Media B. V.

Forbes, K. J. & Gratton, C. (2011). Stable isotopes reveal different patterns of inter-crop dispersal in two ladybeetle species. *Ecological Entomology, 36*, 396–400.

Graham, H. A., Wolfenbarger, D. A. & Nosky, J. B. (1978). Labelling plants and their insect fauna with rubidium. *Environmental Entomology, 7*, 379–383.

Guillebeau, L. P., All, J. N., Nutter, F. W. J. & Kuhn, C. (1993). Comparison of foliar and soil-drench applications of aqueous rubidium chloride solutions to plants for marking feeding aphids (Homoptera: Aphidae). *Journal of Entomological Science, 28*, 370–375.

Hobson, K. A. & Norris, D. R. (2008). Animal migration: A Context for using new techniques and approaches. In: K. A. Hobson & L. I. Wassenaar (Eds.) *Tracking Animal Migration with Stable Isotopes*, vol.2. (pp. 1–19). London: Elsevier, Academic.

Hodek, I. & Michaud, J. P. (2008). Why is *Coccinella septempunctata* so successful? (A point-of-view). *European Journal of Entomology, 105*, 1–12.

Honek, A. & Dixon, A. F. G. & Martinkova, Z. (2008). Body size and the temporal sequence in the reproductive activity of two species of aphidophagous coccinellids exploiting the same resource. *European Journal of Entomology, 105*, 421-425.

Hossain, Z. Gurr, G. M. Wratten, S. D. Raman, A. (2002). Habitat manipulation in lucerne *Medicago sativa*: arthropod population dynamics in harvested and "refuge" crop strips. *Journal of Applied Ecology, 39*, 445–454.

IAEA (2009). *Manual for the use of stable isotopes in entomology*. Vienna.

Johnson, P. C. & Reeves, R. M. (1995). Incorporation of the biological marker rubidium in gypsy moth (Lepidoptera: Lymantriidae) and its transfer to the predator *Carabus nemoralis* (Coleoptera: Carabidae). *Environmental Entomology, 24*, 46–51.

Kleijn, D., Kohler, F., Báldi, A., Batáry, P., Concepción, E. D., Clough, Y., Díaz, M., Gabriel, D., Holzschuh, A., Knop, E., Kovács, A., Marshall, E. J. P., Tscharntke, T. & Verhulst, J. (2009). On the relationship between farmland biodiversity and land-use intensity in Europe. *Proceedings of the Royal Society B: Biological Sciences, 276*, 903–909.

Landis, D. A. Marino, P. C. (1999). Landscape structure and extra-field processes: impact on management of pests and beneficials. In: J. R. Ruberson (Ed.), *Handbook of Pest Management* (pp. 79–109). New York: Marcel Dekker Inc.

Landis, D. A., Wratten, S. D. & Gurr, G. M. (2000). Habitat management to conserve natural enemies of arthropod pests in agriculture. *Annual Review of Entomology, 45*, 175–201.

Lavandero, B., Wratten, S., Hagler, J. & Jervis, M. (2004). The need for effective marking and tracking techniques for monitoring the movements of insect predators and parasitoids. *International Journal of Pest Management, 50*, 147–151.

Lu, Y., Wu, K., Jiang, Y., Guo, Y., Desneux, N. (2012). Widespread adoption of Bt cotton and insecticide decrease promotes biocontrol services. *Nature, 487*, 362–365.

Lumbierres, B., Albajes, R. & Pons, X. (2012). Positive effect of Cry1Ab-expressing Bt maize on the development and reproduction of the predator *Orius majusculus* under laboratory conditions. *Biological Control, 63*, 150-156.

Madeira, F., di Lascio, A., Carlino, P., Costantini, M. L. & Pons, X. (2013). Change in carbon stable isotope ratios of the predatory bug *Orius majusculus* after dietary shifts. *Entomologia Experimentalis et Applicata, 148*, 287–296.

Madeira, F., di Lascio, A., Carlino, P., Costantini, M. L., Rossi, L. & Pons, X. (2014). Stable carbon and nitrogen isotope signatures to determine predator dispersal between alfalfa and maize. *Biological Control, 77,* 66–75.

Madeira, F. & Pons, X. (2015). Effects of weed density on the dispersal of *Orius majusculus* Reuter (Heteroptera Anthocoridae) within maize. *Journal of Applied Entomology, 139*, 712–720.

Madeira, F. & Pons, X. (2016). Rubidium marking reveals different patterns of movement in four ground beetle species between adjacent alfalfa and maize. *Agricultural and Forest Entomology, 18*, 99-107.

Núñez, E. (2002). *La alfalfa como reservorio de enemigos naturales*. PhD Thesis. Universitat de Lleida.

Núñez, E., Rodriguez, E. & Perdiguer, A. (2007). Una alternativa para el control de plagas en la alfalfa: franjas sin cortat como refugio para la fauna auxiliar. *Surcos de Aragón, 101*, 26-30.

Oelbermann, K. & Scheu, S. (2002). Effects of prey type and mixed diets on survival, growth and development of a generalist predator, *Pardosa lugubris* (Araneae: Lycosidae). *Basic and Applied Ecology, 3*, 285–291.

Ostrom, P. H., Colunga-Garcia, M., Gage, S. H. (1997). Establishing pathways of energy flow for insect predators using stable isotope ratios: field and laboratory evidence. *Oecologia 109*, 108–113.

Ouyang, F., Men, X., Yang, B., Su, J., Zhang, Y., Zhao, Z. & Ge, F. (2012). Maize benefits the predatory beetle, *Propylea japonica* (Thunberg), to provide potential to enhance biological control for aphids in cotton. *PloS one, 7*, e44379.

Perović, D. J., Gurr, G. M., Simmons, T. A., Raman, A. (2011). Rubidium labelling demonstrates movement of predators from native vegetation to cotton. *Biocontrol Science and Technology, 10*, 1143–1146.

Pons, X. & Albajes, R. (2002). Control of maize pests with imidacloprid seed dressing treatment in Catalonia (NE Iberian Peninsula) under traditional crop conditions. *Crop Protection, 21*, 943–950.

Pons, X. & Eizaguirre, M. (2009). Cultivos extensivos en regadìo: cereales, maíz y alfalfa. In: J. A. Jacas, A. Urbaneja (Eds.) *Control Biològico de Plagas Agricolas* (pp. 384–398). Valencia, España: PHYTOMA.

Pons, X., Núñez, E., Lumbierres, B. & Albajes, R. (2005). Epigeal aphidophagous predators and the role of alfalfa as a reservoir of aphid predators for arable crops. *European Journal of Entomology, 102*, 519–525.

Pons, X., Lumbierres, B. & Albajes, R. (2009). Heteropterans as aphid predators in inter-mountain alfalfa. *European Journal of Entomology, 106*, 369–378.

Pons, X., Lumbierres, B., Comas, J., Madeira, F. & Starý, P. (2013). Effects of surrounding landscape on parasitism of alfalfa aphids in an IPM crop system in northern Catalonia. *BioControl*, *58*, 733–744.

Post, D. M. (2002). Using stable isotopes to estimate trophic position: models, methods, and assumptions. *Ecology*, *83*, 703–718.

Prasifka, J. R. & Heinz, K. (2004). The use of C3 and C4 plants to study natural enemy movement and ecology, and its application to pest management. *International Journal of Pest Management, 50*, 177–181.

Prasifka, J. R., Heinz, K. M. & Sansone, C. G. (2001). Field testing rubidium marking for quantifying intercrop movement of predatory arthropods. *Environmental Entomology, 30*, 711–719.

Prasifka, J. R., Heinz, K. M. & Winemiller, K. O. (2004). Crop colonisation, feeding, and reproduction by the predatory beetle, *Hippodamia convergens*, as indicated by stable carbon isotope analysis. *Ecological Entomology, 29*, 226–233.

Reynolds, D. R., Riley, J. R., Armes, N. J., Cooter, R. J., Tucker, M. R. & Colvin, J. (1997). Techniques for quantifying insect migration. In: D. R. Dent & M. P. Walton (Eds.) *Methods in Ecological and Agricultural Entomology* (pp. 111–145). Wallingford: CAB International.

Ribera, I., Dolédec, S., Downie, I. S. & Foster, G. N. (2001). Effect of land disturbance and stress on species traits of ground beetle assemblages. *Ecology, 82*, 1112–1129.

Ritchie, S. W., Hanway, J. J. & Benson, G. O. (2005). *How corn plant develops*. Special report No. 48. Iowa State University of Science and Technology. Cooperative Extension Service, Ames, Iowa. p 21.

Scarratt, S., Wratten, S. & Shishehbor, P. (2008). Measuring parasitoid movement from floral resources in a vineyard. *Biological Control, 46*, 107–113.

Schallhart, N., Tusch, M. J., Staudacher, K., Wallinger, C. & Traugott, M. (2011). Stable isotope analysis reveals whether soil-living elaterid larvae move between agricultural crops. *Soil Biology and Biochemistry, 43*, 1612–1614.

Scherber, C., Lavandero, B., Meyer, K. M., Perovic, D., Visser, U., Wiegand, K. & Tscharntke, T. (2012). Scale effects in biodiversity and biological control: Methods and statistical analysis. In: G. M. Gurr, S. D. Wratten, W. E. Snyder & D. M. Y. Read (Eds.) *Biodiversity and Insect Pests: Key Issues for Sustainable Management* (pp. 121–138). Oxford: John Wiley.

Stimmann, M. W. (1974). Marking insects with rubidium: Imported cabbageworm marked in the field. *Environmental Entomology*, *3*, 327–328.

Summers, C. G. (1998). Integrated pest management in forage alfalfa. *Integrated Pest Management Reviews, 154,* 127–154.

Thorbek, P. & Bilde, T. (2004). Reduced numbers of generalist arthropod predators after crop management. *Journal of Applied Ecology, 41*, 526–538.

Tillman, P. G., Prasifka, J. R. & Heinz, K. M. (2007). Rubidium marking to detect dispersal of pest and predator from corn into sorghum and cotton in Georgia. *Journal of Entomological Science*, *42*, 383–391.

Traugott, M., Schallhart, N., Kaufmann, R. & Juen, A. (2008). The feeding ecology of elaterid larvae in central European arable land: New perspectives based on naturally occurring stable isotopes. *Soil Biology and Biochemistry, 40*, 342–349.

Tscharntke, T., Klein, A-M., Kruess, A., Steffan-Dewenter, I., & Thies, C. (2005). Landscape perspectives on agricultural intensification and biodiversity – ecosystem service management. *Ecology Letters, 8*, 857–874.

Tscharntke, T., Bommarco, R., Clough, Y., Crist, T. O., Kleijn, D., Rand, T. A., Tylianakis, J. M., Van Nouhuys, S. & Vidal, S. (2007). Conservation biological control and enemy diversity on a landscape scale. *Biological Control*, *43*, 294–309.

Veres, A., Tóth, F., Kiss, J., Fetykó, K., Orosz, S., Lavigne, C., Otto, S. & Bohan, D. (2012). Spatio-temporal dynamics of *Orius* spp. (Heteroptera: Anthocoridae) abundance in the agricultural landscape. *Agriculture, Ecosystems and Environment, 162*, 45–51.

Vialatte, A., Simon, J. C., Dedryver, C. A., Fabre, F. & Plantegenest, M. (2006). Tracing individual movements of aphids reveals preferential routes of population transfers in agroecosystems. *Ecological Applications, 16*, 839–44.

Wissinger, S. (1997). Cyclic colonization in predictably ephemeral habitats: A template for biological control in annual crop systems. *Biological Control, 10*, 4–15.

Wolfenbarger, D. A., Graham, H. M., Nosky, J. B. & Lindig, H. O. (1982). Boll weevil (Coleoptera: Curculionidae): Marking with rubidium chloride sprays on cotton and dispersal from cotton. *Journal of Economic Entomology, 75*, 1038–1041.

BIOGRAPHICAL SKETCHES

Filipe Madeira, PhD
University of Lleida

Research and Professional Experience:

During my PhD research I was focused on understanding the movement patterns of natural enemies of insect pests in field crop systems (maize, alfalfa and winter cereals). To trace the movement, I mainly used two techniques, rubidium marking and stable isotope analysis of carbon and nitrogen. We were able to show that the dispersal patterns differ between predator species, during the season as well as the fact that the role of each crop as a donor of predators can change during the season. During my graduate studies, I also gained knowledge and experience in both field and laboratory experiments, integrated pest management, insect sampling methods, insect identification, insect and landscape ecology, land management, ecological infrastructures as well as statistical analysis and techniques to understand movement patterns of insects. Furthermore, I collaborated with different international research teams and was responsible for coordinating graduate students in laboratory and field works.

Currently, I am involved in two research topics. The first is to study the influence of the landscape on the predator and prey distribution at different landscape composition: from homogeneous zones with only one (stone fruit orchards or field crops) to zones with mixed landscapes. The second is to work with biological control agents for novel integrated measures in order to reach an effective control of tomato pests.

Professional Appointments:

February 2015 - present

Post-doctoral researcher at the Entomology laboratory of the Department of Crop and Forest Sciences.

University of Lleida, Spain
Advisor: Prof. Dr. Ramon Albajes

May - September 2014
Researcher at the Entomology laboratory of the Department of Crop and Forest Sciences.
University of Lleida, Spain
Advisor: Prof. Dr. Xavier Pons
Education

2009 – 2014
Ph.D., Entomology laboratory of the Department of Crop and Forest Sciences.
University of Lleida, Spain
Dissertation: "Movement of predators in arable crop systems"
Supervisor: Prof. Dr. Xavier Pons

2009 – 2011
M.Sc, Integrated Crop Protection
University of Lleida, Spain
Dissertation: "Dispersal of *Orius majusculus* Reuter (Heteroptera Anthocoridae) in maize with different herbicide treatments"
Supervisor: Prof. Dr. Xavier Pons

2002 – 2008
Graduated in Agricultural and Environmental Engineering
Instituto Politécnico de Castelo Branco, Escola Superior Agrária, Portugal
Dissertation: "Susceptibility of *Pseudaletia unipuncta* Haworth (Lepidoptera Noctuidae) to transgenic and non transgenic maize varieties"
Supervisor: Prof. Dra. Matilde Eizaguirre (UdL), Prof. Dr. Xavier Pons (UdL) and Prof. José Coutinho (EsaCB)

Education at other institutions
October 2012 - December 2012
Period of 2 months (Oct-Dec 2012) in the Faculty of Biology, Department of Invertebrate Zoology and Entomology at the University of Belgrade, involved in research activities and take some training in parasitoid taxonomy, geometric morphometrics and molecular markers in parasitoid identification.
Supervised by Prof. Dr. Željko Tomanović

February 2012 - April 2012

Period of 3 months (Feb- April 2012) in the Agroecology group in the Department of Crop Science at the Georg-August-University Göttingen, to perform a statistic analysis of field data.

Supervised by Dr. Péter Bátary and Prof. Dr. Teja Tscharntke.

April 2011 - September 2011

Period of 5 months (April-Sept. 2011) in the Agroecology group in the Department of Crop Science at the Georg-August-University Göttingen, to carry out a field experiment, and learn new techniques for insect sampling.

Supervised by Dr. Péter Bátary and Prof. Dr. Teja Tscharntke.

September 2010 - December 2010

Period of 3 months (Sept-Dec. 2010) in the Environmental Biology Department at the University of Rome La Sapienza, for learning a methodology for stable isotope analysis of insect samples from maize and alfalfa crops.

Supervised by Prof. Dr. Loreto Rossi

Honors:

November 2015

Extraordinary Doctoral Award in Agricultural sciences – University of Lleida

May 2014

PhD Degree with European Honours

Publications Last Three Years:

1. di Lascio, A., Madeira, F., Carlino, P., Costantini, M.L., Rossi, L. Pons, X. 2016. Movement of three aphidophagous ladybird species between alfalfa and maize revealed by carbon and nitrogen stable isotope analysis. *BioControl* 61:35–46.
2. Madeira and Pons 2016. Rubidium marking reveals different patterns of movement in four ground beetle species between adjacent alfalfa and maize. *Agricultural and Forest Entomology* 18:99-107.
3. Madeira, F. and Pons, X. 2015. Effects of maize weed density on the dispersal of *Orius majusculus* Reuter (Het., Anthocoridae) within maize. *Journal of Applied Entomology* 139:712-720.

4. Madeira F, di Lascio A, Carlino P, Costantini ML, Rossi L, Pons X. 2014. Stable carbon and nitrogen isotope signatures to determine predator dispersal between alfalfa and maize. *Biological Control* 77: 66-75.
5. Madeira F; Di Lascio A; Carlino P; Costantini ML; Pons X. 2013. Change in carbon stable isoyope ratio of the predatory bug *Orius majusculus* after dietary shifts. *Entomologia Experimentalis et Applicata* 148: 287-296.
6. Pons X; Lumbierres B; Comas J; Madeira F; Starý P. 2013. Effects of surrounding landscape on parasitism of alfalfa aphids in an IPM crop system in northern Catalonia. *BioControl* 58: 733-744.
7. Albajes,R., Lumbierres,B., Madeira, F., Comas, C., Ardanuy, A., Marina Lee, M., Iglesias, S., Comas, J., Pérez-Hedo, M., López, C., Eizaguirre, M., Pons, X. 2013 Field trials for assessing risks of GM maize on non-target arthropods in Europe: the Spanish experience. IOBC/WPRS. Bulletin 97:1-8.

Xavier Pons, PhD
University of Lleida

Research and Professional Experience:

I have developed my scientific career in the framework of the Entomology applied to the pest control, especially aphids, in arable crops in Mediterranean conditions. The research lines have been focused on the integrated pest management in winter cereals, alfalfa, and especially maize. In this last crop, I have been mainly working in determining and evaluating the effect of transgenic maize resistant to lepidopterans (Bt Maize) on non-target arthropods. Since my PhD graduation, I have been continuously participating in several projects from the Plan Nacional de I+D+I and I have been the Principal Researcher from one of them. The most important features achieved have been: (1) the knowledge, in arable crops like winter cereals, alfalfa and maize, of the composition and relative importance of pest species and natural enemies, their biology and their phenology; (2) the damages caused by these pests and the control methods, including mainly natural enemies; (3) to determine that Bt maize do not produce significant changes on the composition and abundance of non-target herbivores and their natural enemies, and that the Bt-maize use is compatible with programmes of integrated pest management; (4) to determine and to evaluate the movement of predators between crops, and at farm scale, and the derived effects in the maintenance of populations in order to enhance conservation biological control, and (5) to establish methodologies of integrated management for the control of alfalfa pests. Most of the scientific articles listed in point C.1. come from this research lines. The acquired experience has allowed signing several contracts with international companies and local cooperatives in which I have participated as a researcher or as principal researcher.

Since 2000, in addition to the above the described research lines, I have been the leader of a new line related with the integrated pest management in urban green areas. Up today, this line has been funded by private contracts with companies or by councils of several cities. Results related to the species identification (especially aphids and their natural enemies), the relative importance of the pests, the development of sampling methods and decision making systems adapted to the urban environment. is demanded for companies, city councils and other corporations involved in pest control in the urban

Professional Appointments

From 1986 to now: Assistant professor, Associated professor, Lecturer and Full Professor at the University of Lleida

Education

1986 Ph.D., Universidad Politécnica de Catalunya

1983 Ingeniero Agrónomo, Universidad Politécnica de Catalunya

1978 Ingeniero Técnico Agrícola, Universidad Politécnica de Catalunya

Publications Last Three Years:

Di Lascio A, Madeira F, Costantini ML, Rossi L, Pons X. 2016. Movement of three aphidophagous ladybird species between alfalfa and maize revealed by carbon and nitrogen stable isotope analysis. *BioControl* 61: 35-46.

Madeira and Pons 2016. Rubidium marking reveals different patterns of movement in four ground beetle species between adjacent alfalfa and maize. *Agricultural and Forest Entomology* 18:99-107.

Pons X, Roca M, Lumbierres B, Lucas E. 2015. Characterization of a newly established aggregation of the invasive ladybeetle *Harmonia axyridis* and current status of the invader in Spain. *Spanish Journal of Agricultural Research* 13 (2) 21006 11 pages. Doi 10.5224/sjar/2015132-7054.

Comas, J.; Lumbierres, B.; Comas, C.; Pons, X.; Albajes, R. 2015. Optimising the capacity of field trials to detect the effect of genetically modified maize on non-target organisms through longitudinal sampling. *Annals of Applied Biology* 166:183-195.

Lumbierres B, Pérez Hidalgo N, Starý P, Pons X. 2015. First record of the invasive *Siphonatrophia cupressi* (Swain) (Aphididae: Aphidini: Aphidina) in the Iberian peninsula.: *Redia-Giornale Di Zoologia* 98: 141-144.

Madeira F, Pons X. 2015. Effects of weed density on the dispersal of *Orius majusculus* Reuter (Heteroptera: Anthocoridae) within maize. *Journal of Applied Entomology* 139:712-720.

Madeira F, di Lascio A, Carlino P, Costantini ML, Rossi L, Pons X. 2014. Stable carbon and nitrogen isotope signatures to determine predator dispersal between alfalfa and maize. *Biological Control* 77: 66-75.

Comas C, Lumbierres B, Pons X, Albajes R. 2014. No effects of *Bacillus thuringiensis* maize on nontarget organisms in the field in southern Europe: a meta-analysis of 26 arthropod taxa. *Transgenic Research* 23: 135-143

Albajes, R.; Lumbierres, B., Pons, X; y Comas, J. 2014. Changes in arthropod fauna from weed management practices in genetically modified herbicide-tolerant maize. *Journal of Agricultural Science* 6: 67-78.

Comas J, Lumbierres B, Pons X, Albajes R.2014. Ex-Ante Determination of the Capacity of Field Tests to Detect Effects of Genetically Modified Corn on Nontarget Arthropods. *Journal of Economic Entomology* 106: 1659-1668.

Havelka J, Tomanovic Z, Kos K, Kavallieratos N, Janecek J, Pons X, Rakhshani E, Stary P. 2014. Mountain aphid and parasitoid guilds on Aconitum spp. in Europe. *Bulletin of Insectology* 67: 57-61.

Romeis J, Meissle M, Alvárez-Alfageme F, Bigler F, Bohan DA, Devos Y, Malone LA, Pons X, Rauschen S. 2014. Potential use of an arthropod database to support the non-target risk assessment and monitoring of transgenic plants. *Transgenic Research* DOI 10.1007/s11248-014-9791-2

Pons X; Lumbierres B; Comas J; Madeira F; Starý P. 2013. Effects of sorrounding landscape on parasitism of alfalfa aphids in an IPM crop system in northern Catalonia. *BioControl* 58: 733-744.

Madeira F; Di Lascio A; Carlino P; Costantini ML; Pons X. 2013. Change in carbon stable isotope ratio of the predatory bug *Orius majusculus* after dietary shifts. *Entomologia Experimentalis et Applicata* 148: 287-296.

Albajes R, Lumbierres B, Pons X, Comas J. 2013. Representative taxa in field trials for environmental risk assessment of genetically modified maize. *Bulletin of Entomological Research* 103: 724-733.

Vasileiadis VP; Moonen AC; Sattin M; Otto S; Pons X; Kudsk P; Veres A; Dorner Z; van der Weide R; Marraccini E; Pelzer E; Angevin F; Kiss J. 2013. Sustainability of European maize-based cropping systems: Economic, environmental and social assessment of current and proposed innovative IPM-based systems. *European Journal of Agronomy* 48: 1-11.

In: Natural Enemies
Editor: Sónia A.P. Santos
ISBN: 978-1-63485-921-9

Chapter 5

BIOECOLOGY OF *PSYTTALIA CONCOLOR* (SZÉPLIGETI) (HYMENOPTERA: BRACONIDAE) AND MANAGEMENT STRATEGIES TO INCREASE BIOLOGICAL CONTROL OF THE OLIVE FRUIT FLY

***Lara A. Pinheiro*[1,2], *Pilar Medina*[3], *Albino Bento*[1], *José Alberto Pereira*[4] and *Sónia A. P. Santos*[1,5,*]**

[1]Mountain Research Center (CIMO), School of Agriculture, Polytechnic Institute of Bragança, Bragança, Portugal
[2]Escuela Superior y Técnica de Ingeniería Agraria, Universidad de León, León, Spain
[3]Unidad de Protección de Cultivos, E.T.S.I. Agrónomos, Universidad Politécnica de Madrid, Madrid, Spain
[4]School of Agriculture, Polytechnic Institute of Bragança, Bragança, Portugal
[5]Barreiro School of Technology, Polytechnic Institute of Setúbal, Lavradio, Portugal

[*] Corresponding author: Sónia A. P. Santos. Mountain Research Center, School of Agriculture, Polytechnic Institute of Bragança, Campus de Santa Apolónia, 5300-253 Bragança, Portugal. Email: saps@ipb.pt.

Abstract

Olive growing is an important land-use in the Mediterranean region, with significant environmental, social and economic roles. However, the olive tree is attacked by several pests that reduce the yield, the olive oil quality and growers' income. Among those pests, the olive fruit fly, *Bactrocera oleae* (Rossi) (Diptera: Tephritidae), is the major pest of commercial olives worldwide. In sustainable olive groves, the control of the olive fruit fly should be achieved by the maximization of the action of its natural enemies. Among the complex of natural enemies, *Psyttalia concolor* (Szépligeti) (Hymenoptera: Braconidae) has been reported as an important endoparasitoid of the olive fruit fly as well as of other tephritids of economic importance. However, conservation or enhancement of *P. concolor* populations in agroecosystems hinges on the maintenance of non-crop plants (e.g., vegetation strips and bushes) within or around the olive groves since adult parasitoids require floral resources (pollen and nectar) or honeydew to fulfill their energetic needs.

Other key issues are the influence of temperature, olive tree cultivars, pesticides or competition with other species of parasitoids on *P. concolor* performance. In this chapter, the bioecology of *P. concolor* and management strategies required to increase biological control of the olive fruit fly in olive orchards will be reviewed and new insights will be discussed concerning the manipulation of ecological infrastructures.

Keywords: *Bactrocera oleae*, biological control, *Olea europaea*, parasitoid, *Psyttalia concolor*

1. Introduction

The olive tree, *Olea europaea* L. is an emblematic plant in Mediterranean countries where it represents an important social, economic and environmental role. Olive groves are complex agroecosystems harboring organisms belonging to different trophic levels, some of them causing economic losses such as pathogens (e.g., *Verticillium dahliae* Kleb and *Colletotrichum acutatum* Simmonds) or phytophagous insects [e.g., *Bactrocera oleae* (Rossi) (Diptera: Tephitidae) and *Prays oleae* (Bernard) (Lepidoptera: Praydidae)] whereas others exert limiting action on these olive enemies such as entomophagous predators [e.g., *Chrysoperla carnea* (Stephens) and *Chilocorus bipustulatus* (Linnaeus)], parasitoids such as *Psyttalia lounsburyi* (Silvestri), *Ageniaspis fuscicollis* (Dalman) or *Coccophagus lycimnia* (Walker) and pathogen antagonists such as *Bacillus subtilis* (Ehrenberg) (Daane et al., 2015; Ghribi et al., 2011; López-Escudero & Mercado-Blanco, 2011; Mnif & Ghribi, 2015; Paredes et al., 2015; Santos et al., 2009).

Among pests, the olive fruit fly *B. oleae* is a monophagous frugivorous pest feeding on *Olea* species (i.e., cultivated *O. europaea* and wild relatives) (Nardi et al., 2005), and is considered the key pest with higher economic impact on olive production and product quality (table olives and olive oil) (Malheiro et al., 2015). The economic losses are related with the consumption of the fruit pulp by the newly hatched larvae, normally ranging from 50 to 150 mg/larvae (Neuenschwander & Michelakis, 1978). Moreover, one female can oviposit 200-250 eggs during its lifetime (Mavragani-Tsipidou, 2002).

Control methods traditionally used against the olive fruit fly rely on the insecticide application. However, good agricultural practices (GAP) have become an important issue in agricultural systems as a way to ensure environmental sustainability and food safety and quality, which is not compatible with the excessive use of insecticides. Therefore, an alternative agronomic practice to conventional management is the use of natural enemies reported to attack the olive fruit fly. Some of the parasitoid species referred are the ectoparasitoids *Eupelmus urozonus* Dalman (Hymenoptera: Eupelmidae) and *Pnigalio mediterraneus* Walker (Hymenoptera: Eulophidae) and endoparasitoids such as *Psyttalia concolor* (Szépigeti) and *P. lounsburyi* (Hymenoptera: Braconidae) (Daane et al., 2015; Daane & Johnson, 2010).

Historically, *P. concolor* was almost the only exotic parasitoid used in European biological control programs against the olive fruit fly, already tested in inoculative and inundative mass releases (Raspi & Loni, 1994), or combined with other techniques, such as mass trapping (Hepdurgun et al., 2009; Liaropoulos et al., 2002; 2005). This species was originally described as *Opius concolor* by Szépligeti (1910), being the first description based on samples collected from olives infested with olive fruit flies in Tunisia. Afterwards, *P. concolor* was introduced in Italy to control the olive fruit fly (Delucchi, 1957) being successively used in biocontrol programs.

This parasitoid can be easily reared in laboratory on the Mediterranean fruit fly, *Ceratitis capitata* (Wiedemann) (Diptera: Tephritidae), not requiring great costs unlike other parasitoids (González-Núñez, 1998). In addition, another advantage of the use of *P. concolor* in biological control programs is its great capacity to survive to shipping conditions during importation process (Yokoyama et al., 2006). In the field, the success of a parasitoid species depends on its survival, dispersal, attack rate and multi-generational persistence (Sivinski & Aluja, 2012).

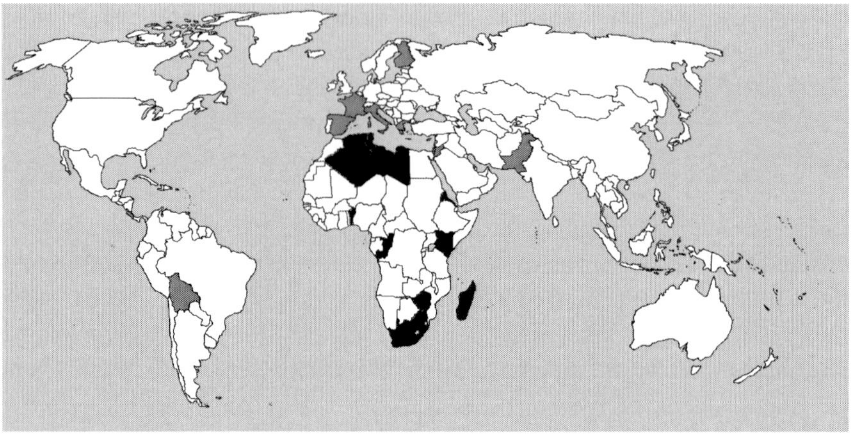

Figure 1. Geographical distribution of *Psyttalia concolor*: In black are represented its native countries and in gray are represented the countries where it was introduced.

Besides attacking a target pest, a parasitoid should also not attack other important species in the ecosystem (namely non-pestiferous congeners). In fact, it is known that *P. concolor* attacks at least 14 tephritid pests (Diptera) in its native range and in areas where it has been introduced (Canale et al., 2014). Thus, for an effective pest control, several factors affecting the success of the parasitoid need to be clarified. In this chapter, the biology, reproduction, behavior, ecology of the parasitoid and components of the interaction *P. concolor*-olive fruit fly are described and new insights to the conservation of the parasitoid in olive groves are discussed.

2. Distribution of *Psyttalia concolor*

Psyttalia genus has 79 valid species (Wharton, 2009). Among them, *P. concolor* is native to Africa (Figure 1) but its taxonomic situation is confusing, with some authors associating this species to a species complex and others dealing with it as synonyms. For a more detailed description of taxonomic problems associated with *P. concolor* see Kimani-Njogu et al. (2001), Karam et al. (2008) and Rugman-Jones et al. (2009).

In the last decades, several efforts were attempted to introduce *P. concolor* in a number of countries (Figure 1), allowing its spread, in order to control the population of the olive fruit fly and of other pests (e.g., *C. capitata*). Several releases aiming the biological control of the olive fruit fly by *P. concolor* occurred in Italy (Monastero, 1968; Monastero & Delanoue, 1966), Greece (Kapatos et al., 1977), Spain (Jimenez et al., 1990; Miranda et al., 2008), and California (Yokoyama et al., 2006; 2008). Although numerous releases were attempted, only few were successful, such as those described by Raspi & Loni (1994) in Italy and Miranda et al. (2008) in Majorca - Balearic Islands (Spain). Several factors affected the effectiveness of *P. concolor* in olive groves, namely the longevity (in the presence or absence of host) and the accessibility of food source. In the present chapter these factors are exposed in detail. It is of great importance to create strategies that will contribute to an adaptation and establishment of *P. concolor* under different environmental conditions.

3. Biology and Life-History of *Psyttalia concolor*

3.1. The Life Cycle of *Psyttalia concolor*

P. concolor is a synovigenic koinobiont larval-pupal endoparasitoid (Figure 2) that fully depends on the host for its development until the adult stage. *P. concolor* life-cycle has four developmental stages: egg, larva (with four stages: one initial, two intermediates and one terminal), pupa (red-eye pupa or young pupa, and black-eyed pupa or old pupa) and adult (González et al., 1996). *P. syttalia concolor* is a parasitoid of second- and third-instar larvae of tephritids, although it might also parasitize freshly molted pupae. The life cycle begins when a *P. concolor* adult female lays its eggs onto the host larvae, such as the olive fruit fly (Figure 2), feeding the immature parasitoid on tissues (or hemolymph) of the host body (Harvey & Strand, 2002). Therefore, the fitness of larvae is related to the quality of the host and host selection decision by the oviposition female.

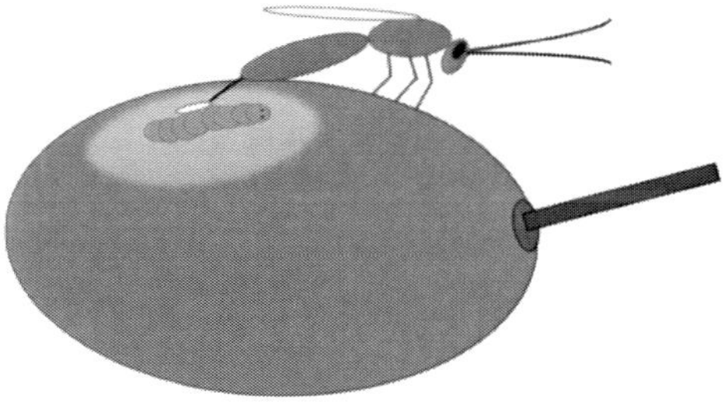

Figure 2. *Psyttalia concolor* laying one egg on an olive fruit fly larvae.

Normally, *P. concolor* life-cycle takes around 17.87 to 21.52 (at 23 and 25°C, respectively) from egg to the new adult born, with the eclosion taking around 6-7 minutes (Biliotti & Delanoue, 1959; Loni, 1997). At 24-25°C, egg incubation takes generally one day and larval and pupal stages take about 5-9 days and 6-15 days, respectively (Bartlett et al., 1978).

3.2. Longevity of *Psyttalia concolor* Adults

P. concolor males emerge between 2-3 days before females at 25 ± 2°C, i.e., 18-19 days after host parasitization. Females have a higher survival rate than males, living about 30 and 17 days (respectively) under laboratory conditions (25 ± 2°C) (González-Núñez, 1998). This reveals a sexual discrepancy in developmental time probably due to sexual dimorphism. Thus, a faster development of males might represent an increment in mating success. In females, potential fecundity and oviposition rates are correlated with body size (Berger et al., 2012), likely representing an investment in reproduction potential that ensures future survival of the specie.

In the field, longevity is influenced by climatic conditions (temperature and relative humidity), availability of food sources and costs of sexual interaction. On the other hand, according to Yokoyama et al. (2006), longevity of *P. concolor* adults does not depend on the presence of infested and non-infested olive fruits, which only increases the parasitism rate and the stability of the parasitoid population. Sime et al. (2006) also showed that *P. concolor* life span decreases significantly when food resources are scarce, even in the presence of hosts. Moreover, in greenhouse experiments, adult parasitoids without food and water lived 4 and 1 days, in simulated coastal climate and simulated inland valley climate, respectively (Yokoyama et al., 2008). These results demonstrate that the parasitoid can live longer in a colder and more humid climate than in a warmer and drier environment and the presence of water and food sources is essential and increases survival in the agroecosystems.

Costs of sexual interactions (courtship, mating and reproduction) are generally assumed to be high both for males (that include gamete production, courtship and combat and scramble competition) and females (that include physical damage caused by male genitalia, toxic seminal fluid and offspring production and care) (Benelli et al., 2013a). For both sexes, the presence of other males and females interferes on their longevity. Male longevity is extremely reduced when they are kept with females or other males, being female longevity reduced in the presence of males comparing with the presence of females (Benelli et al., 2013a). High rates of progeny were produced when parasitoids were about 12-16 days old (Sime et al., 2006; Yokoyama et al., 2006).

3.3. Courtship and Mating Behavior

Courtship and mating behavior are the keys for reproduction and knowledge of these parameters is extremely important for the success of mass rearing and optimization of biological control programs. Responsiveness and mating success were affected by the periodicity and duration of mating, being longer and more successful in the early morning than in the afternoon. However, in males, mating is not affected by the female mating status (virgin or mated females) (Benelli et al., 2012). Some authors justify that mated females continue to produce sex attaints after mating; or that the presence of visual signals emitted by the female are sufficient to stimulate a sexual response from males (Benelli et al., 2012; De Freitas et al., 2004). Concerning *Aphidius ervi* (Hymenoptera: Braconidae) McClure et al. (2007) divided the courtship phase into three steps: (1) a *courtship phase*, where male uses its antennas to interact with the female through visual and chemical cues; (2) followed by a *close-range phase*, getting females attention through wing fanning; and lastly (3), a *contact phase*, characterized by acoustic, mechanical and tactile movements of the antennas leading to mating efforts. This sequence of courtship is similar to the one described for *P. concolor*, with the courtship song being emitted at a frequency of 180 Hz (Benelli et al., 2012; Canale et al., 2013). In experiments conducted by Benelli et al. (2012), *P. concolor* spent 20.3 ± 3.49 seconds in pre-copulatory courtship and the male approached the female with one or more (till a maximum of nine) copulation attempts.

Another interesting aspect concerning *P. concolor* courtship is related with male sexual experience. Since a male-male interaction was observed before female emergence, Benelli and Canale (2012a; 2013) hypothesized that this previous mating experience would improve mating intensity. However, no differences were observed in mating success and copula duration.

3.4. Host-Stage Preference

The interaction host-parasitoid is also an important factor in biological control programs optimization, specifically in inundative/augmentative releases. Females spend a large part of its life looking for potential hosts and the success of parasitoid development depends on parameters such as female age, size of the host, host exposure to females and learning experience.

P. concolor is able to discriminate the host instar, searching preferentially for the second and third instars of the olive fruit fly, but revealing an increase in the reproduction success when larvae are in the third instar (Canale & Loni, 2006; Raspi & Canale, 2000; Sime et al., 2006; Yokoyama et al., 2008). Canale and Loni (2006) showed the difficulties of *P. concolor* in locating young host larvae. According to those authors, the parasitoid identifies the larvae inside the fruit through vibrations emitted by the host, being easier to detect higher vibrations produced by the third instar larvae. On the other hand, Kouame and Mackauer (1991) suggested that the preference for a specific host stage could also be the result of the ability of larvae to defend itself from the parasitoid, being the cost of lost "opportunity time" lesser in larvae with a weaker immune response.

The size of the host larvae influences the parasitoid size, with greater host larvae originating an increase in the offspring, wing size and the number of flagellomeres (antennal segments excluding scape and pedicel) (Billah et al., 2005). In the same study, the relationships between host size, host type and tibia lengths were analyzed, suggesting that the size of the tibia increased with the host size. However, differences in the ovipositor-tibia ratio and ovipositor sheath-tibia ratio were not found when *P. concolor* was reared on bigger hosts.

In other experiments, when large hosts were exposed to *P. concolor* females, the sex ratio was independent of female age (Avilla & Albajes, 1984). However, the sex ratio was influenced by the size and number of hosts exposed to *P. concolor* females, emerging more males when pupae were small but in few numbers (Avilla & Albajes, 1984; Loni, 2003). This can be explained due to the fact that *P. concolor* females mate only a single time (monandrous females) and, contrastingly, the males mate multiple times, not requiring many males in an offspring to fertilize all the females (Ragusa, 1974). *P. concolor* seems to have a preference for non-parasitized larvae. However, if larvae have already been parasitized, the parasitoid preferred hosts with fewer parasitizations due to competition, resulting in a lower return in offspring number and quality (Benelli, Gennari & Canale, 2013). However, and still according to the same authors, *P. concolor* was not able to recognize the author of larvae parasitization (self-parasitized vs. conspecífic-parasitized).

Furthermore, antennas are used as chemoreceptors (particularly the last antennomeres) and play an important role in a first host location behaviour, being the ovipositor important for exact host location and host discrimination (Barbarossa et al., 1998; Canale & Raspi, 2000). After larvae approach, the oviposition is usually successful after 30-40 seconds (Canale & Raspi, 2000). In addition to size, fruit characteristics such as epicarp and pulp depth and firmness, and seed size and morphology can also influence the accessibility to the host larvae (Sivinski, 1991; 2001).

3.5. Visual, Olfactory and Mechanical Stimuli

The success of *P. concolor* in finding one suitable tephritid host in the orchard (habitat) or fruit (micro-habitat location) is the result of olfactory (volatile semiochemicals), visual and/or mechanical cues (Benelli et al., 2013b; Quilici & Rousse, 2012). There are some techniques (e.g., olfactometry and/or electroantennography) that allow studying the influence of these factors on the ovipositor-probing response (Quilici & Rousse, 2012). Acording to Stuhl et al. (2011), volatile organic compounds (VOCs) emitted by plants (from leaves, flowers or fruits) and/or by tephritid larvae, while feeding on the olive pulp, elicit a response in antennae and ovipositor sensillas.

Benelli et al. (2013b) reported that volatile compounds of healthy and infested fruits may represent a first approximation to the host selection. Subsequently, VOCs derived from short-range micro-habitat (e.g., derived from substrate fermentation) and/or excretion (by the larvae) help in host location.

Color might also have an important role in *P. concolor* host searching behavior. Benelli and Canale (2012b) demonstrated a color preference through laboratory trials, associating larvae with a given color (after a previous training period). The authors also showed a reaction to color hue, but not to differences in light intensity. However, no preference for shapes and patterns were found. Transposing to field conditions, *P. concolor* might use previous location experiences to reduce searching and decision time for hosts.

Once *P. concolor* females are on the surface of the fruit, sounds or vibrations emitted by the feeding larvae are crucial for host location. Canale (2003) also demonstrated that interaction between mobility of larvae and host substrate elicited an ovipositor-probing behavior due to a change of the chemical composition of parasitized fruits, when compared with healthy fruits. In a situation of larvae immobilization, other factors such as body temperature and CO_2 radiation are recognized by the female, leading to an oviposition behavior (Meyhöfer & Casas, 1999).

Beyond chemical, visual, and mechanical stimuli, factors such as learning, climatic conditions and physiological state of the parasitoid influence host location (Quilici & Rousse, 2012). During host search, data on female parasitoid response indicate that localization and oviposition behaviors are synergistic.

3.6. Life-Time Fecundity

P. concolor female is a synovigenic species (i.e., depends on nutrients for egg maturation) and each female produces around 60 mature eggs (Genduso, 1969). The egg laying distribution pattern is associated with finding suitable or unsuitable larvae (Loni, 2003). If *P. concolor* female accepts the host, lays one egg on each parasitization (Canale & Loni, 2006). However, Loni (2003) described that the highest percentage of host larvae are superparasitized with two eggs in order to ensure the developed of the parasitoid. Zhang & Raspi (1999) also demonstrated that self-superparasitism occurred at a parasitoid: host ratio of 1:5 using both "experienced" and "inexperienced" females, after an exposure period of 6 h. Superparasitism refers to the oviposition behavior of female when lays an egg in a host already parasitized and can be a strategy to maximize the progeny development by reducing host larvae immune defense system. In fact, the venom injected during the oviposition of each egg may be insufficient to weaken the immune-system, in particular of the latest host instar larvae (Loni, 2003; Raspi & Canale, 2000). However, only one larva remains alive after superparasitism.

The lifetime production of offspring observed per *P. concolor* female varied from 22.2 to 28.7 days with higher rates of progeny being produced during the first 12-14 days under controlled conditions (22 ± 2°C, 40% RH fed on a solution of honey - 50%) (Sime et al., 2006). Adult females of parasitoids search for nectar and honeydew during the first few days of life instead of hosts (Jervis & Heimpel, 2005). When food resources are limited and females do not have nutrients for somatic maintenance, they obtain them by reabsorbing the eggs (Collier, 1995). This suggests that sources of nutrition are essential for reproduction success, as adult performance involves high cost of energy and nutrient accumulation, especially by the females. In this sense, a good strategy for a successful biological control program would be to increase the potential fecundity and fertility with e.g., supplemental pollen, since this will enhance *P. concolor* population (Russell et al., 2015).

4. Ecology and Habitat Relationship

4.1. *Psyttalia concolor* as a Biological Control Agent of *Bactrocera oleae*

The role of parasitoid-pest relationship in ecosystem functioning refers to the importance of *P. concolor* in limiting *B. oleae* emergence through the parasitization of infested olive fruits (Figure 3). Oviposition success and pest control can be evaluated through the parasitism rate that shows the impact of the parasitoid on the pest population, being this parameter studied in olive orchards for many years. In 1965, in Halki Island (Greece), 25,000 individuals of *P. concolor* were released and the parasitoid was successfully established, reaching parasitism rates of 50% at the end of the study (Stavraki-Pavlopoulou, 1967). In another release, in Corfu, Kapatos et al. (1977) observed that 30-50% of the host larvae were parasitized one week after the release. In Jaen (Spain), after three consecutive years of releases, individuals of *P. concolor* were collected from olives, showing the winter permanence of the parasitoid (Jimenez et al., 1990).

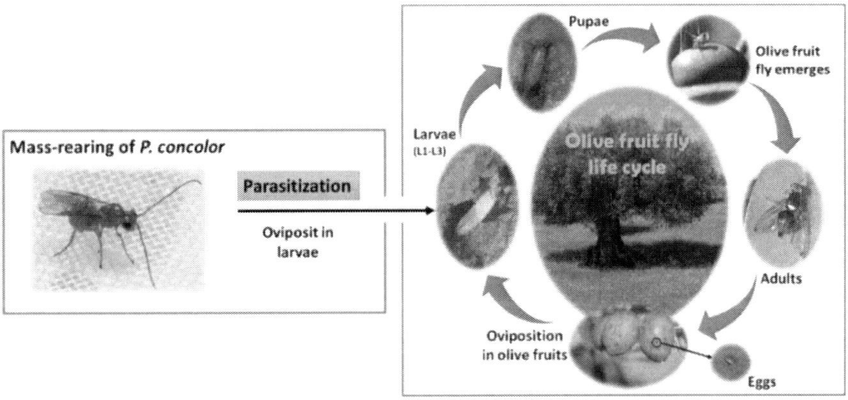

Figure 3. Interaction between *Psyttalia concolor* female and *Bactrocera oleae*.

One of the most recent studies, reported in two organic olive orchards located in Majorca (Balearic Islands, Spain), parasitism rates of *P. concolor* in *B. oleae* ranged from 6.0% to 40.0% with seasonal averages of 22.4% and 23.4% in each orchard (Miranda et al., 2008). Also, *P. concolor* releases performed in San José (coastal location) and Grapevine Road (inland valley location), California, lead to parasitism rates from 0.5 to 4.0% and 1.5 to 30.0%, respectively (Yokoyama et al., 2006).

Mathé-Hubert et al. (2013) estimated parasitism rates through PCR method for *P. concolor* and demonstrated an increase of parasitism rates with the time of exposure of 1 h and 8 h from 12.0% to 52.0%, respectively. However, *P. concolor* parasitism seems to be low when compared to other parasitoids. Other studies considering factors that limit the maintenance of the parasitoid in olive orchards need to be performed.

4.2. Thermal Compatibility between *Psyttalia concolor* and *Bactrocera oleae*

Temperature is an extremely strong factor delimiting survival, reproduction, emergence development and behavior of insects (e.g., feeding, migration) (González et al., 1996; Mohamed et al., 2006). Loni (1997) reported a mean development time of *P. concolor* ranging between 14.62 and 77.25 days for temperatures varying from 15ºC to 30ºC. Under extreme hot temperatures (>30°C) or cold temperatures (<13°C), *P. concolor* development is restricted and the reproduction activity is practically inhibited and often lethal (Loni, 1997), while *B. oleae* develops and survives from 10 to 30°C (Wang et al., 2012). When compared with *B. oleae*, *P. concolor* presented a similar low threshold (10.01ºC and 11.70ºC, respectively) (Loni, 1997). This statement shows a positive match concerning biological thermal requirements. Therefore, temperature not only affects behavior and performance of *P. concolor* but also survival of their host species and might affect the success of biocontrol programs.

At optimal temperature (20-25°C), females quickly mate (5-10 seconds) after emergence and host parasitization highly increase (Loni, 1997). When temperatures reach 27°C, *P. concolor* initiates the thermal stress (Loni, 1997). A curious fact is that *P. concolor* cannot complete any developmental stage at 10ºC, but stays alive during 23 to 29 days as larva, 31 days as pupa, becoming lethal for eggs after 6 days (Genduso, 1970). González et al. (1996) observed a similar lethality on eggs, but a different conclusion was reported for the last larval stage that survived longer, registering then a higher emergence than the first and second larval stages. Besides that, no losses were found in the emergence beneficial capacity in pupae with 14 days (González et al., 1996). These results suggested that the inability of a parasitoid species for establishing during biological control programs may be attributed to many factors beyond the temperature, such as, inhospitable environments or seasonal gaps in alternative host availability or specific food resources (e.g., pollen or nectar of spontaneous flowers).

4.3. Effect of Pesticides on *Psyttalia concolor*

Insecticides are management tools intended to minimize the impact of pests on farmer's crops and economies. Concerning the olive fruit fly, traditional control of this pest has been based in organophosphate insecticides (e.g., dimethoate), pyrethroids (e.g., lambda-cyhalothrin) or spinosad (Skouras et al., 2007; Margaritopoulos et al., 2008). Although the efficacy of these insecticides in short-term mortality is a point to consider, others like ecological (e.g., degradation of residues through time) and toxicological (non-target species can be directly and indirectly affected) effects on natural enemies of pests should also be evaluated (Guedes et al., 2016). The toxicity of an insecticide depends on its class, as well as the species and developmental stage of the insect. Some of these toxicological side effects in natural enemies are: (1) acute toxicity, driving to mortality or (2) chronic toxicity, causing changes in reproduction (e.g., number of eggs), behavior performance (e.g., parasitization activity), biochemical or physiological effects (e.g., changes in enzyme activity or respiration) (Amano & Haseeb, 2001). A number of studies showed the effect of pesticides on longevity and reproduction of *P. concolor* (e.g., Adán et al., 2007; Bengochea et al., 2012; González et al., 1998a; 1998b; Jacas & Viñuela, 1994; Medina et al., 2006, 2008; Sterk et al. 1999, Viñuela et al., 2000, 2001; Zapata et al., 2005). Thus, a significant decrease of the emergence of adults was observed when *P. concolor* pupae were treated with dimethoate; but treatments with kaolin, Bordeaux mixture and copper oxychloride did not affected emergence (Bengochea et al., 2014). The mortality rate of females was also studied after 72 h of exposure to pesticides applied on glass surfaces, on olive tree leaves and under semi-field conditions. Similarly to the previous experiment, only dimethoate showed deleterious effect, with 100% of mortality after 24 h, except on the semi-field experiment (where 100% of mortality was observed after 72 h). However, reproduction was not significantly affected (Bengochea et al., 2014).

Other experiments showed that the ingestion of azadirachtin caused a negative impact on longevity, number of attacked hosts and progeny size of P. *concolor* females, with these parameters showing a decrease of 84.4, 15.6 and 32.5% respectively, when compared with the control treatment (Viñuela et al., 2000). More studies concerning the effect of pesticides on *P. concolor* are needed, which allow to harmonize the application of pesticides and the maintenance of *P. concolor* in the olive grove.

4.4. Food Sources for *Psyttalia concolor* – a Case Study

The landscape structure complexity (e.g., occurrence of flower strips or woody plants) can contribute to reducing *B. oleae* population in olive groves and increasing parasitoid community (Ortega & Pascual, 2014) (Figure 4).

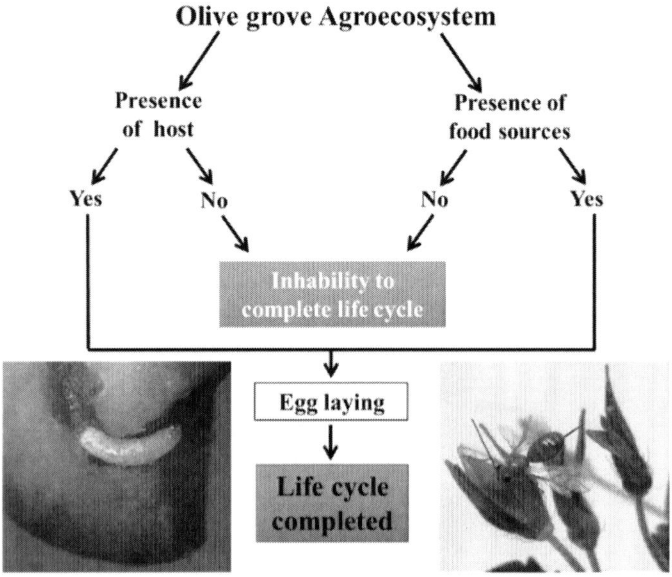

Figure 4. Interactions between *Psyttalia concolor* and components of the olive grove agroecosystem.

Parasitoids consume different anatomical parts of plants and chemical signaling emitted by plants are recognized as a source of food and/or hosts. Parasitoids feed (a) directly on pollen and nectar (floral and extrafloral) and, in less quantity, on trichomes and epidermis or (b) indirectly, through honeydew produced by Hemiptera species (Jervis et al., 1993).

According to Wäckers (2005), parameters that might affect food suitability are: (1) *Availability:* food sources can vary in time and space (e.g., phenological stages, temperatures); (2) *Apparency*: detection by visual (color, shape, movement) and chemical cues (olfaction, gustation); (3) *Accessibility*: location of food source, mouthpart morphology of the insect and the characteristics of the food source and (4) *Composition/nutritional suitability*: chemical nature and proportion of nutrients needed for survival and reproduction (i.e., the level and type of carbohydrates, amino acids, proteins, lipids and secondary metabolites).

In particular, spontaneous or introduced flowering plants provide nutrients and energy and, as a result, improve longevity and fecundity of adult parasitoids, contributing to maintaining them in the agroecosystem (Balzan & Wäckers, 2013; Begum et al., 2006; Russell, 2015). Fitness and success are determined by plant-Hymenoptera interaction which involves, for example, the compatibility between the corolla opening and parasitoid mouthparts (Jervis et al., 1993). Sivinski et al. (2011) showed that floral area in the field and corolla width significantly influenced the abundance of Braconidae. Furtado et al. (in press) evaluated the effect of several plant species on longevity of *P. concolor* by offering the extracted nectar or the flowering plant. *Rosmarinus officinalis* L. (Lamiaceae) positively affected *P. concolor* longevity when its nectar was provided, while *Foeniculum vulgare* L. (Apiaceae) caused a significant increase on *P. concolor* longevity when flowers (containing nectar) were given. Interestingly, *Anchusa azurea* Mill. (Boraginaceae) and *Lavatera cretica* L. (Malvaceae) caused high longevity, disregarding nectar accessibility. Further than showing a preference of the parasitoid for shallow corollas, these results suggested that flowers had potential for maximising the fitness of the parasitoid in olive groves. Moreover, the production of honeydew (sugar solution) by the black-scale, *Saissetia oleae* (Olivier) (Hemiptera: Coccidae), the olive psyllid, *Euphyllura olivina* (Costa) (Hemiptera: Psyllidae) and aphids on wildflowers are alternative nutritional resources that can complement parasitoid diet. According to Furtado et al. (in press), honeydew excreted by *Aphis gossypii* Glover (Hemiptera: Aphididae) and *E. olivina* equally increased the longevity of *P. concolor* (14.27 and 13.67 days respectively).

Also, the occurrence of alternate hosts that guarantee the conservation of *P. concolor* (when *B. oleae* larvae is not present) throughout the year needs further research to complement food sources studies. According to Wharton (2009), *Psyttalia* species can be reared on fruits such as *Psyttalia masneri* Wharton that was reared on fruits of *Dracaena fragrans* (L.) Ker Gawl. (Liliaceae) infested with *Taomyia marshalli* Bezzi (Diptera: Tephritidae) and *Psyttalia halidayi* Wharton on fruits of *Lettowianthus stellatus* Diels (Annonaceae) infested with *Ceratitis rosa* Karsch (Diptera: Tephritidae). *P. concolor* was obtained from hosts of *Ceratitis cosyra* (Walker) reared on fruits of *Icacina senegalensis* (A) Juss. (Ica.) (Icacinaceae) in Casamance (Senegal) in 2010 (Vayssières et al., 2012). Asteraceae is almost exclusively associated with *Tephritinae* subfamily (Farrell et al., 1992) and it was referred as a potential source of alternate *P. concolor* hosts in olive groves (Cobo et al., 2014). However, other plant families (Cucurbitaceae, Myrtaceae, Rosaceae or Rubiaceae) are considered sources of tephritid flies (Copeland et al., 2002; Hernandez-Ortiz & Perez-Alonso, 1993; Leblanc et al., 2012).

Below, a case study is presented in order to show the effect of different sugars and amino acids, commonly present in pollen, nectar and/or honeydew, on longevity of females and males and beneficial capacity of *P. concolor*.

Methods: *P. concolor* used in the study were obtained from a mass rearing culture on *C. capitata* maintained at the Crop Protection Unit of the Technical University of Madrid, Spain.

The food sources tested were (1) water (negative control), (2) a mixture of hydrolysed protein and sugar (1:4) (solid diet - positive control), (3) glucose, (4) fructose, (5) sucrose, (6) melibiose, (7) trehalose, (8) melezitose, (9) sorbitol, (10) a mixture of glucose and phenylalanine (Phe), (11) a mixture of glucose and proline (Pro), (12) a mixture of glucose and tryptophan (Trp), (13) a mixture of glucose, Phe, Pro and Trp. The concentration of sugar solutions was 1 M and amino acids 0.1 mM. Water was also provided in all treatments. The food sources tested were offered in a glass vial (2 ml) and the solid diet was placed on a plastic stopper cover ad libitum.

For the longevity and parasitization experiments, cylindrical acrylic boxes (12 cm height and 5 cm in diameter) with a hole on the top (5.5 cm in diameter) covered with a piece of permeable cloth for ventilation. For the longevity experiments, five newly emerged couples (<24 h) were placed in each box and a total of 30 males and 30 females were tested with one of the diets mentioned above. The mortality of *P. concolor* was daily recorded. For parasitization experiments, one mated female per replicate was maintained isolated in a cage, feeding during five days after its emergence with one of the diets tested. Then, 20 third instar *C. capitata* larvae were offered for 2 hours to each female, repeated during five consecutive days (the first day was removed from data analyses), using a cage previously described by Bengochea et al. (2014). A total of 30 females per amino acid and 25 females per sugar were tested. The influence of the different food sources on the beneficial capacity of *P. concolor* was measured as the percentage of hosts attacked (i.e., without medfly emergence), progeny size and parasitoid emergence. The experiments were performed in a climatic chamber at T = 25 ± 2ºC; HR = 75 ± 5% and 16:8 h L:D photoperiod. Statistical analyses were performed with IBM-SPSS statistics, version 19.0.0 (SPSS Inc. IBM Company, 2010). The normal distribution of the residuals and the homogeneity of variance were evaluated by means of the Kolmogorov-Smirnov test and the Levene´s tests, respectively. Longevity and the parameters of beneficial capacity were conducted for comparison among treatments by using analysis of variance (ANOVA) followed by the Tukey - Kramer HSD test. All statistical tests were performed at a 5% significance level. In addition, for every treatment, longevity and parasitoid emergence was compared between males and females following an independent t-test.

Results and discussion: Individuals fed on glucose plus amino acid solutions showed that females lived significantly longer than males when fed on the different food treatments. When fed on water, the effect was the opposite, suggesting that females were more susceptible to the lack of food than males. These results also indicated that mixtures of amino acid and glucose increased *P. concolor* longevity, specially of females. However, there is no synergistic effect between amino acids since the mixture of the three amino acids with glucose did not increase longevity (for both females and males). The effect of phenylalanine and proline consumption on *P. concolor* longevity was significantly higher than water and glucose. However, no significant differences were found when compared with tryptophan and mixture (Phe + Pro + Trp) (Table 1).

The different sugars had a significant effect on the longevity of *P. concolor* when compared with individuals fed on water only. Differences on longevity between treatments are higher for females than males, which reinforces the idea that the quality of nutrition is more decisive for the females than for males (Table 2). Results indicate that a diet rich in amino acids and carbohydrates is especially important for survival of *P. concolor* females. The effect of phenylalanine, melibiose and sorbitol increased the longevity (Table 1 and Table 2) and further investigations should potentially study the effect of consumption of these three resources on the performance of the species.

Furthermore, it is important to confirm that *P. concolor* is able to use a particular dietary source other than purely ingesting it.

In the parasitization experiments, a high reduction of the percentage of progeny of individuals feed on glucose + phenylalanine was found. Also, a greater variation between females emerged between treatments was noticed. The percentage of attacked hosts was significantly lower for individuals fed on glucose + proline than for those fed on glucose + phenylalanine, glucose only and the mixture of glucose with the other amino acids (Table 3).

When melezitose was offered, the reproductive parameters were not different from those observed on water (Table 4). Considering sucrose, an increase on the percentage of females emerged was observed but did not show significant differences from the positive control (Table 4).

These results suggest that the consumption of amino acids can be important for egg maturation, potentially leading to higher parasitism rates and parasitoid emergences, and carbohydrates offered isolated did not favor reproduction. Pointing this, it is important to understand which nutrients are associated with reproduction and future research should focus on the determination of sugars in the hemolymph and fat body reserves of *P. concolor*.

Table 1. Longevity in days (mean ± standard error of the mean - SE) of *Psyttalia concolor* for each tested amino acid + glucose, mixture of amino acids + glucose, water only (negative control) and glucose (positive control). n = 30 individuals/treatment for females and males. The concentration used was 0.1 mM for the amino acids and 1 M for glucose

Food source	Longevity (days) (Mean ± SE)			
	Female	Male	P^1	Total
Water only	2.23 ± 0.08^a	3.47 ± 0.09^a	***	2.85 ± 0.10^a
Glucose	43.87 ± 3.05^b	29.73 ± 3.18^b	**	36.80 ± 2.37^b
Glucose + Phenylalanine	59.43 ± 1.31^c	31.10 ± 1.97^b	***	45.27 ± 2.18^{cd}
Glucose + Proline	56.07 ± 2.58^c	30.60 ± 2.00^b	***	43.33 ± 2.32^{bd}
Glucose + Tryptophan	52.47 ± 3.21^{bc}	26.63 ± 1.80^b	***	39.55 ± 2.48^{bd}
Glucose + Phe + Pro + Trp	49.73 ± 2.79^{bc}	31.83 ± 2.12^b	***	40.78± 2.09^{bd}

Mean (± SE) followed by the same letter on the same column are not significantly different at P > 0.05; [1] shows the results for comparison between males and females where n.s. not significant; $^{*}p < 0.05$; $^{**}p < 0.01$; $^{***}p < 0.001$

Table 2. Longevity in days (mean ± standard error of the mean - SE) of *Psyttalia concolor* for each tested sugar, solid diet (positive control) and water only (negative control). n = 30 individuals/treatment for females and males. The concentration used for sugar solutions was 1 M

Food source	Longevity (days) (Mean ± SE)			
	Female	Male	P[1]	Total
Water only	1.90 ± 0.07^{a}	3.23 ± 0.15^{a}	***	2.57 ± 0.12^{a}
Solid diet	48.87 ± 4.25^{bd}	26.33 ± 2.03^{bd}	***	37.60 ± 2.76^{b}
Sugars Monosaccharides				
Fructose	35.07 ± 3.57^{def}	23.23 ± 1.72^{bd}	***	29.15 ± 2.11^{bd}
Glucose	40.13 ± 4.29^{dg}	26.77 ± 2.14^{bd}	***	33.45 ± 2.53^{bd}
Disaccharides				
Melibiose	57.57 ± 1.43^{b}	13.20 ± 1.91^{c}	***	35.38 ± 3.12^{bd}
Sucrose	17.27 ± 2.83^{c}	19.17 ± 1.89^{b}	***	18.22 ± 1.69^{cd}
Trehalose	20.40 ± 1.69^{ce}	21.83 ± 1.05^{bd}	***	21.12 ± 0.99^{bd}
Trisaccharides				
Melezitose	23.37 ± 2.71^{c}	19.23 ± 1.74^{b}	***	21.30 ± 1.62^{bd}
Sugar alcohol				
Sorbitol	52.20 ± 4.87^{bfg}	32.17 ± 4.06^{d}	***	42.18 ± 3.40^{b}

Mean (± SE) with the same letter in the column are not significantly different at $P > 0.05$: [1] shows the results for comparison between males and females where n.s. not significant; $^{*}p < 0.05$; $^{**}p < 0.01$; $^{***}p < 0.001$

Table 3. Percentage of attacked hosts and progeny and parasitoid emergence for 4 days of parasitization (mean ± standard error of the mean - SE) for each amino acid + glucose, mixture of amino acids + glucose, water only (negative control) and glucose (positive control) tested. n = 30 females (1 *Psyttalia concolor*/20 larvae of *Ceratitis capitata*)

Food source	Attacked hosts (%)	Progeny (%)	Parasitoid emergence (Mean ± SE)		
			Female	Male	P[1]
Water only	53.2 ± 6.6^{a}	51.1 ± 4.9^{a}	19.6 ± 2.7^{a}	6.2 ± 1.1^{a}	***
Glucose	92.1 ± 1.6^{bd}	78.9 ± 1.3^{b}	41.3 ± 2.4^{bd}	17.0 ± 2.0^{b}	***
Glucose + Phenylalanine	90.6 ± 2.6^{bd}	14.9 ± 1.1^{c}	37.7 ± 1.9^{bd}	11.0 ± 0.8^{ab}	***
Glucose + Proline	75.9 ± 4.7^{ce}	61.8 ± 3.9^{ad}	24.2 ± 3.2^{ac}	15.9 ± 2.8^{b}	n.s.
Glucose + Tryptophan	89.7 ± 2.1^{bde}	64.6 ± 2.6^{de}	32.9 ± 2.4^{cd}	13.9 ± 1.4^{ab}	***
Glucose + Phe + Pro + Trp	92.8 ± 2.0^{b}	76.2 ± 2.4^{be}	40.7 ± 2.8^{d}	16.0 ± 2.5^{b}	***

Note: Mean (± SE) with the same letter in the column are not significantly different at $P > 0.05$; [1] shows the results for comparison between males and females where n.s. not significant; $^{*}p < 0.05$; $^{**}p < 0.01$; $^{***}p < 0.001$

Table 4. Percentage of attacked hosts and progeny and parasitoid emergence for 4 days of parasitization (mean ± standard error of the mean - SE) for each tested sugar, solid diet (positive control) and water only (negative control). n = 25 females (1 *Psyttalia concolor*/20 larvae of *Ceratitis capitata*)

Food source	% Attacked hosts	% Progeny	Parasitoid emergence (Mean ± SE)		
			Female	Male	P^1
Water only	53.2 ± 6.6[a]	51.1 ± 4.9[ad]	19.6 ± 2.7[ad]	6.2 ± 1.1[a]	***
Solid diet	89.9 ± 2.3[b]	74.5 ± 2.5[bc]	33.4 ± 4.1[bc]	20.5 ± 3.4[b]	*
Sugars (1 M)					
Monosaccharides					
Fructose	92.1 ± 1.6[b]	63.3 ± 1.5[ac]	34.4 ± 2.1[bc]	12.5 ± 1.8[ab]	***
Glucose	90.6 ± 1.6[b]	62.9 ± 1.5[acd]	27.6 ± 3.4[abcd]	18.0 ± 3.3[bc]	n.s.
Disaccharides					
Melibiose	83.3 ± 3.8[b]	52.2 ± 3.4[ad]	24.5 ± 2.6[abd]	12.4 ± 2.2[ab]	***
Sucrose	94.9 ± 1.2[b]	64.6 ± 1.6[ac]	37.8 ± 1.7[c]	11.5 ± 1.2[ab]	***
Trehalose	85.1 ± 3.4[b]	62.6 ± 2.6[acd]	22.7 ± 3.4[abd]	20.8 ± 3.6[b]	n.s.
Trisaccharides					
Melezitose	48.9 ± 6.5[a]	48.4 ± 6.5[d]	17.8 ± 3.0[d]	7.5 ± 1.4[acd]	**
Sugar alcohol					
Sorbitol	85.2 ± 2.1[b]	62.4 ± 1.6[acd]	24.4 ± 3.1[abd]	18.5 ± 2.9[bd]	n.s.

Note: Mean (± SE) with the same letter in the column are not significantly different at $P > 0.05$.; [1] shows the results for comparison between males and females where n.s. not significant; * $p < 0.05$; ** $p < 0.01$; *** $p < 0.001$

CONCLUSION

This chapter focused the main factors that influence the abundance of *P. concolor* in the olive grove. Food sources have an impact on the parasitoid longevity, but their influence on pest/parasitoid interaction is still unknown, as well as how they affect biological control and, consequently, management decisions.

The establishment of *P. concolor* in a new environment depends on a combination of abiotic and biotic factors. Some factors such as fruit size, weight, color, fruit epicarp hardness, phenological stage of the olive fruit and chemical factors may influence the attractiveness of olive cultivars to *P. concolor* female. It is clear that landscape structure plays an important role on parasitoid survival, nutrition and activity. Thus, it is imperative to study *P.*

concolor feeding habits and nutritional requirements, and their consequences to longevity, development and reproduction. The abundance of alternative hosts it is equally important, especially in the absence of olive fruits. Pointing this, it is essential to find plants with alternative tephritid host flies in olive groves and conserve or enhance the abundance of these plants.

Moreover, it is essential to understand host - *P. concolor* mechanisms, as well as host - associated stimuli and response behavior of females (e.g., antennal search, ovipositor search and vibrotaxis). Many questions about the interaction between *P. concolor* and *B. oleae* still need to be clarified. It is also important to identify the volatile compounds involved in these interactions, which can influence host identification and oviposition behavior of *P. concolor*.

These studies will contribute to promote the best management practices in the olive grove and provision of biological control of the olive fruit fly by attracting and conserving *P. concolor* in the crop. Moreover, future research about increasing *P. concolor* longevity leading to higher parasitism rates and lower *B. oleae* densities should be exploited. Other natural enemies of the olive fruit fly should also be surveyed as agents that might be used with *P. concolor*.

ACKNOWLEDGMENTS

The authors are grateful to the Portuguese Foundation of Science and Technology (FCT) for financial support through the projects EXCL/AGR-PRO/0591/2012 "Olive crop protection in sustainable production under global climatic changes: linking ecological infrastructures to ecosystem functions." L.A. Pinheiro is grateful to FCT for the PhD studentship (SFRH/BD/103998/2014); S.A.P. Santos is grateful to FCT for the Post-doctoral fellowship (SFRH/BPD/102277/2014). This chapter is part of L.A. Pinheiro PhD Thesis.

REFERENCES

Adán, A., González, T., Bastante, R., Budia, F., Medina, P., Del Estal, P. & Viñuela, E. (2007). Efectos de diversos insecticidas aplicados en condiciones de laboratorio extendido sobre *Psyttalia concolor* (Szépligeti) (Hymenoptera: Braconidae) [Effects of various insecticides applied in laboratory conditions extended *Psyttalia concolor* (Szépligeti) (Hymenoptera: Braconidae)]. *Boletín de Sanidad Vegetal. Plagas*, 33, 391-397.

Amano, H. & Haseeb, M. (2001). Recently-proposed methods and concepts of testing the effects of pesticides on the beneficial mite and insect species: study limitations and implications in IPM. *Journal Applied Entomology*, 36, 1-11.

Avilla, J. & Albajes, R. (1984). The influence of female age and host size on the sex ratio of the parasitoid *Opius concolor. Entomologia Experimentalis et Applicata*, 35, 43-47.

Balzan, M.V. & Wackers, F.L. (2013). Flowers to selectively enhance the fitness of a host-feeding parasitoid: adult feeding by *Tuta absoluta* and its parasitoid *Necremnus artynes*. *Biological Control*, 67, 21-31.

Barbarossa, I.T., Muroni, P., Dardani, M., Casula, P. & Angioy, A.M. (1998). New insight into the antennal chemosensory function of *Opius concolor* (Hymenoptera: Braconidae). *Italian Journal of Zoology*, 65, 367-370.

Bartlett, B.R., Clausen, C.P., DeBach, P., Goeden, R.D., Legner, E.F., McMurtry, J.A. & Oatman, E.R. (1978). In: C.P. Clausen (Ed.). *Introduced parasites and predators of arthropod pests and weeds: A world review* (pp. 329-331). Washington: Department of Agriculture, Agricultural Research Service.

Begum, M., Gurr, G., Wratten, S.D., Hedberg P.R. & Nico, H.I. (2006). Using selective food plants to maximize biological control of vineyard pests. *Journal of Applied Ecology*, 43, 547-554.

Benelli, G., Bonsignori, G., Stefanini, C. & Canale, A. (2012). Courtship and mating behaviour in the fruit fly parasitoid *Psyttalia concolor* (Szépligeti) (Hymenoptera: Braconidae): the role of wing fanning. *Journal of Pest Science*, 85, 55-63.

Benelli, G & Canale, A. (2012a). Do *Psyttalia concolor* (Hymenoptera: Braconidae) males gain in mating competitiveness from being courted by other males while still young? *Entomological Science*, 15, 257-260.

Benelli, G & Canale, A. (2012b). Learning of visual cues in the fruit fly parasitoid *Psyttalia concolor* (Szépligeti) (Hymenoptera: Braconidae). *BioControl*, 57, 767-777.

Benelli, G. & Canale, A. (2013). Male-male sexual behavior in the parasitic wasp *Psyttalia concolor*. *Journal of Insect Science*, 13:25.

Benelli, G., Gennari, G. & Canale, A. (2013). Host discrimination ability in the tephritid parasitoid *Psyttalia concolor* (Hymenoptera: Braconidae). *Journal of Pest Science*, 86, 245-251.

Benelli, G., Gennari, G., Francini, A. and Canale, A. (2013a). Longevity costs of same-sex interactions: first evidence from a parasitic wasp. *Invertebrate Biology*, 132, 156-162.

Benelli, G., Revadi, S., Carpita, A., Giunti, G., Raspi, A., Anfora, G. & Canale, A. (2013b). Behavioral and electrophysiological responses of the parasitic wasp *Psyttalia concolor* (Szépligeti) (Hymenoptera: Braconidae) to *Ceratitis capitata*-induced fruit volatiles. *Biological Control*, 64, 116-124.

Bengochea, P., Christiaens, O., Amor, F., Viñuela, E., Rougé, P., Medina, P. & Smagghe, G. (2012). Ecdysteroid receptor docking suggests that dibenzoylhydrazine-based insecticides are devoid to any deleterious effect on the parasitic wasp *Psyttalia concolor* (Hym. Braconidae). *Pest Management Science*, 68, 976-985.

Bengochea, P., Budia, F., Viñuela, E. & Medina, P. (2014). Are kaolin and copper treatments safe to the olive fruit fly parasitoid *Psyttalia concolor*? *Journal of Pest Science*, 87, 351-359.

Berger, D., Olofsson, M., Friberg, M. Karlsson, B., Wiklund, C. & Gotthard, K. (2012). Intraspecific variation in body size and the rate of reproduction in female insects - adaptive allometry or biophysical constraint? *Journal of Animal Ecology*, 8, 1244-1258.

Biliotti, E. & Delanoue, P. (1959). Contribution a l'étude biologique d' *Opius concolor* Szépl. (Hym. Braconidae) en élevage de laboratoire [Contribution to the biological study of *Opius concolor* Szépl. (Hym. Braconidae) in laboratory culture]. *Entomophaga*, 4, 7-14.

Billah, M.K., Kimani-Njogu, S., Overholt, W.A., Wharton, R.A., Wilson D.D. & Cobblah, M.A. (2005). The effect of host larvae on three *Psyttalia* species (Hymenoptera: Braconidae), parasitoids of fruit-infesting flies (Diptera: Tephritidae). *International Journal of Tropical Insect Science*, 25, 168-175.

Canale, A. (2003). *Psyttalia concolor* (Hymenoptera Braconidae): role of host movement and host substrate in ovipositor-probing behaviour. *Bulletin of Insectology*, 56, 211-213.

Canale, A. Benelli, G., Lanzo, F., Giannotti, P. Mazzoni, V. & Lucchi, A. (2013). The courtship song of fanning males in the fruit fly parasitoid *Psyttalia concolor* (Szépligeti) (Hymenoptera: Braconidae). *Bulletin of Entomological Research*, 103, 303-309.

Canale, A., Geri, S. & Benelli, G. (2014). Associative learning for host-induced fruit volatiles in *Psyttalia concolor* (Hymenoptera: Braconidae), a koinobiont parasitoid of tephritid flies. *Bulletin of Entomological Research*, 104, 774-780.

Canale, A. & Loni, A. (2006). Host location and acceptance in *Psyttalia concolor*: role of host instar. *Bulletin of Insectology*, 59, 7-10.

Canale, A. & Raspi, A. (2000). Host location and oviposition behaviour in *Opius concolor* Szépligeti (Hymenoptera, Braconidae). *Entomological Problems*, 31, 25-32.

Cobo, A., González-Núñez, M., Sánchez-Ramos, I. & Pascual, S. (2014). Selection of non-target tephritids for risk evaluation in classical biocontrol programmes against the olive fruit fly. *Journal Applied Entomology*, 139, 179-191.

Collier, T.R. (1995). Host Feeding, Egg Maturation, Resorption, and Longevity in the Parasitoid *Aphytis melinus* (Hymenoptera: Aphelinidae). *Annals of the Entomological Society of America*, 88, 206-214.

Copeland, R.S., Wharton, R.A., Luke, Q. & De Meyer, M. (2002). Indigenous hosts of *Ceratitis capitata* (Diptera: Tephritidae) in Kenya. *Annals of the Entomological Society of America*, 95, 672-694.

Daane, K.M. & Johnson, M.W. (2010). Olive Fruit Fly: Managing an Ancient Pest in Modern Times. *Annual Review of Entomology*, 55, 151-169.

Daane, K.M., Wang, X., Nieto, D.J., Pickett, C.H., Hoelmer, K.A., Blanchet, A. & Johnson, M.W. (2015). Classic biological control of olive fruit fly in California, UA: release and recovery of introduced parasitoids. *BioControl*, 60, 317-330.

De Freitas, M.R.T., De Lima Medonça, A., Do Nascimento, R.R. & Sant'Ana, E.G. (2004). Behavioural evidence for a female sex pheromone in *Cotesia flavipes* (Hymenoptera: Braconidae). *Physiological Entomology*, 29, 183-187.

Delucchi, V. (1957). Les parasites de la mouche des olives [The parasites of the olive fly]. *Entomophaga*, 2: 107-118.

Farrell, B.D., Mitter, C. & Futuyma, D.J. (1992). Diversification at the insect-plant interface, insights from phylogenetics. *BioScience*, 42, 34-42.

Furtado, C. Belo, A. F., Nunes, F.M., Ganhão, E., Müller, C.T., Torres, L. & Rei, F. T. (in press). Evaluating potential olive orchard sugar food sources for the olive fly parasitoid *Psyttalia concolor*. *BioControl*. DOI: 10.1007/s 10526-016-9732-5.

Genduso, P. (1970). Influenza della temperatura sulla durata e sulla resistenza degli stadi preimmaginali dell' *Opius c. siculus* Mon. [Influence of temperature on the duration and strength of the stadium preimaginal' *Opius c. siculus* Mon.]. *Bolletino dell'Instituto di Entomologia Agraria e dell' Observatorio di Fitopatologia di Palermo* [Bulletin of the Institute of Agricultural Entomology and Observatorio Phytopathology Palermo], 8, 1-7.

Ghribi, D., Mnif, I., Boukedi, H., Kammoun, R., Ellouze-Chaabouni, S., (2011). Statistical optimization of low-cost medium for economical production of *Bacillus subtilis* biosurfactant, a biocontrol agent for the olive moth *Prays oleae*. *African Journal of Microbiology Research*, 5, 4927-4936.

González, M., Bahena, F. & Viñuela, E. (1998a). Efectos de distintos reguladores del crecimiento de insectos (RCI) sobre el parasitoide *Opius concolor* Szépligeti cuando son ingeridos por la larva huésped [Effects of different insect growth regulators (IGR) on the parasitoid *Opius concolor* Szépligeti when ingested by the host larva]. *Boletín de Sanidad Vegetal. Plagas*, 24, 193-200.

González, M., Bahena, F. & Viñuela, E. (1998b). Desarrollo de un método de semicampo para el estudio de los efectos secundarios de los productos fitosanitarios sobre el parasitoide *Opius concolor* Szépligeti [Development of a method semifield to study the side effects of plant protection products on the parasitoid Opius concolor Szépligeti]. *Boletín de Sanidad Vegetal. Plagas*, 24, 661-668.

González, M., Jacas, J.A., Jiménez, A. & Viñuela, E. (1996). Optimización de la cría de *Opius concolor* Szepl. mediante la utilización de bajas temperaturas durante su desarrollo pre-imaginal [Optimization of breeding Opius Szépl concolor using low temperatures during development pre-imaginal]. *Boletín de* Sanidad *Vegetal. Plagas*, 22, 107-114.

González-Núñez, M. (1998). *Uso conjunto de plaguicidas y enemigos naturales en el olivar: Optimización del manejo de* Opius concolor *Szépligeti, parasitoide de la mosca del olivo,* B. oleae *(Gmelin)* [Joint use of pesticides & natural enemies in the olive grove: Optimizing management Opius Szépligeti concolor, parasitoid of the olive fruit fly, *B. oleae* (Gmelin)]. PhDThesis. Madrid: Universidad Politécnica de Madrid. ETSI Agrónomos.

Guedes, R.N.C., Smagghe, G., Stark, J.D & Desneux, N. (2016). Pesticide-induced stress in arthropod pests for optimized integrated pest management programs. *Annual Review of Entomology*, 61, 3.1-3.20.

Harvey J.A. & Stran, M.R. (2002). The developmental strategies of endoparasitoid wasps vary with host feeding ecology. *Ecology*, 83, 2439-2451.

Hepdurgun, B. Turanli, T. & Zümreoğlu, A. (2009). Parasitism rate and sex ratio of *Psyttalia (=Opius) concolor* (Hymenoptera: Braconidae) reared on irradiated *Ceratitis capitata* larvae (Diptera: Tephritidae). *Biocontrol Science and Technology*, 19: 157-165.

Hernández-Ortiz, V. & Pérez-Alonso, R. (1993). The natural host plants of *Anastrepha* (Diptera: Tephritidae) in a tropical rain forest of Mexico. *The Florida Entomologist*, 76, 447-460.

Jacas J.A. & Viñuela, E. (1994). Normalización del método para estudiar los efectos secundarios de los fitosanitarios sobre *Opius concolor* Szépligetti (Hym. Braconidae) [Standardization of the method to study the side effects of crop protection on *Opius concolor* Szépligetti (Hym. Braconidae)]. *Boletín de Sanidad Vegetal. Plagas*, 20, 409-417.

Jervis, M.A. & Heimpel, G.E. (2005). Phytophagy. In: M.A. Jervis (Ed.), *Insects as Natural Enemies: A Practical Perspective* (pp. 525-550). Dordrecht: Springer.

Jervis, M.A., Kidd, N.A.C., Fitton, M.G., Huddleston, T. & Dawah, H.A. (1993). Flower-visiting by hymenopteran parasitoids. *Journal of Natural History*, 27, 67-105.

Jimenez, A., Castillo, E. & Lorite, P. (1990). Supervivencia del himenóptero bracónido *Opius concolor* Szép. parásito de *Dacus oleae* Gmelin. en olivares de Jaén [Survival of the hymenopteran braconid *Opius concolor* Szép. *Dacus oleae* Gmelin parasite in olive groves of Jaén]. *Boletín de Sanidad Vegetal. Plagas*, 16, 97-103.

Kapatos, E., Fletcher, B.S., Pappas, S. & Laudeho, Y. (1977). The release of *Opius concolor* and *O. concolor* (var. *siculus* [Hym.: Braconidae] against the spring generation of *Dacus oleae* [Dipt.: Trypetidae] on Corfu. *Entomophaga*, 22, 265-270.

Karam, N., Guglielmino, C.R., Bertin, S., Gomulski, L.M., Bonomi A., Baldacchino, F., Simeone, V. & Malacrida, A.R. (2008). RAPD analysis in the parasitoid wasp *Psyttalia concolor* reveals Mediterranean population structure and provides SCAR markers. *Biological Control*, 47, 22-27.

Kimani-Njogu, S.W., Trostle, M.K., Wharton, R.A., Woolley, J.B. & Raspi, A. (2001). Biosystematics of the *Psyttalia concolor* species complex (Hymenoptera: Braconidae: Opiinae): the identity of populations attacking *Ceratitis capitata* (Diptera: Tephritidae) in coffee in Kenya. *Biological Control*, 20, 167-174.

Kouame, K.L. & Mackauer, M. (1991). Influence of aphid size, age and behaviour on host choice by the parasitoid wasp *Ephedrus californicus*: a test of host-size models. *Oecologia*, 88, 197-203.

Leblanc, L., Vueti, E. T., Drew, R A. I. & Allwood, A. J. (2012). Host plant records for fruit flies (Diptera: Tephritidae: Dacini) in the Pacific Islands. *Proceedings of the Hawaiian Entomological Society*, 44, 11-53.

Liaropoulos, C., Mavraganis, V.G., Broumas, T. & Ragoussis, N. (2002). Field tests on the efficacy of mass trapping combined with the release of *Opius concolor* (Hymenoptera: Braconidae) parasites for the control of the olive fruit fly *Bactrocera oleae* (Diptera: Tephritidae). 7th European Congress of Entomology, October 7-13, Thessaloniki (Greece).

Liaropoulos, C., Mavraganis, V.G., Broumas, T. & Ragoussis, N. (2005). Field tests on the combination of mass trapping with the release of the parasite *Opius concolor* (Hymenoptera: Braconidae), for the control of the olive fruit fly *Bactrocera oleae* (Diptera: Tephritidae). In: Proceedings of the Working Group on Integrated Protection of Olive Crops, Florence, Italy: *IOBC/WPRS Bulletin*, 30: 77-81.

Lombardo, N. (2003). Aspetti generali dell'olivicoltura [General aspects of olive]. In: P. Fiorino (ed.) *Olea, Tratatto di Olivicoltura* [Olea, Steamed of Olive] (pp. 3-12). Bologna: Ed. Edagricole.

Loni, A. (1997). Developmental rate of *Opius concolor* (Hym.: Braconidae) at various constant temperatures. *Entomophaga*, 42, 359-366.

Loni, A. (2003). Impact of host exposure time on mass-rearing of *Psyttalia concolor* (Hymenoptera Braconidae) on *Ceratitis capitata* (Diptera Tephritidae). *Bulletin of Insectology*, 56, 277-282.

López-Escudero, F.J., & Mercado-Blanco, J. (2011). *Verticillium* wilt of olive: a case study to implement an integrated strategy to control a soil-borne pathogen. *Plant Soil*, 344, 1-50.

Malheiro, R., Casal, S., Baptista, P. & Pereira, J.A. (2015). A review of *Bactrocera oleae* (Rossi) impact in olive products: From the tree to the table. *Trends in Food Science and Technology*, 44, 226-242.

Margaritopoulos, J.T., Skavdis, G., Kalogiannis, N., Nikou, D., Morou, E., Skouras, P.J., Tsitsipis, J.A. & Vontas, J. (2008). Efficacy of the pyrethroid alpha-cypermethrin against *Bactrocera oleae* populations from Greece, and improved diagnostic for an iAChE mutation. *Pest Management Science*, 64, 900-908.

Mathé-Hubert, H., Gatti, J-L., Poirié, M. & Malausa, T. (2013). A PCR-based method for estimating parasitism rates in the olive fly parasitoids *Psyttalia concolor* and *P. lounsburyi* (Hymenoptera: Braconidae). *Biological Control*, 67, 44-50.

Mavragani-Tsipidou, P. (2002). Genetic and cytogenetic analysis of the olive fruit fly *Bactrocera oleae* (Diptera: Tephritidae). *Genetica*, 116, 45-57.

McClure, M., Whistlecraft. J. & McNeil, J.N. (2007). Courtship behaviour in relation to the female sex pheromone in the parasitoid *Aphidius ervi* (Hymenoptera: Braconidae). *Journal of Chemical Ecology*, 33, 1946-1959.

Medina, P., Corrales, E., González-Núñez , M., Smagghe, G. & Viñuela, E. (2008). Effects of *Beauveria bassiana, Heterorhabditis bacteriophora, H. megidis* and *Sterinernema feltiae* on the Mediterranean fruit fly *Ceratitis capitata* and the very sensitive braconid *Psyttalia concolor. IOBC/wprs Bulletin*. 35, 113-121.

Medina, P., Budia, F., Rodriguez, B., González, M., Diaz, A., Huerta, A., Zapata, N. & Viñuela, E. (2006) Effects of botanical insecticides on two natural enemies of importance in Spain: *Chrysoperla carnea* (Stephens) and *Psyttalia concolor* (Szépligeti). *IOBC/wprs Bulletin*. 29, 85-93.

Meyhöfer, R. & Casas, J. (1999). Vibratory stimuli in host location by parasitic wasps. *Journal of Insect Physiology*, 45, 967-971.

Miranda, M.A., Miquel, M., Terrassa, J., Melis, N. & Monerris, M. (2008). Parasitism of *Bactrocera oleae* (Diptera, Tephritidae) by *Psyttalia concolor* (Hymenoptera, Braconidae) in the Balearic Islands (Spain). *Journal of Applied Entomology*, 132, 798-805.

Mnif, I. & Ghribi, G. (2015). Potential of bacterial derived biopesticides in pest management. *Crop Protection*, 77, 52-64.

Mohamed, S.A., Overholt, W.A., Wharton, R.A. & Lux, S.A. (2006). Effect of temperature on developmental time and longevity of *Psyttalia cosyrae* (Hymenoptera: Braconidae). *Biocontrol Science and Technology*, 16, 717-726.

Monastero, S. (1968). Nouvelle expérimentation à grande échelle de la lutte biologique contre la mouche de l'olive (*Dacus oleae* Gmel.) au moyen d' *Opius concolor siculus* Mon. en Sicile en 1967 [New large-scale testing of biological control against the olive fly (*Dacus oleae* Gmel.) by means of *Opius concolor* siculus Mon. Sicily in 1967]. *Entomophaga*, 13, 251-261.

Monastero, S. & Delanoue P. (1966). Lutte biologique expérimentale contre la mouche de l'olive (*Dacus oleae* Gmel.) au moyen d' *Opius concolor* Szépl. *siculus* Mon. dans les iles éoliennes (Sicile) en 1965 [experimental biological control against the olive fly (*Dacus oleae* Gmel.) by means of *Opius concolor* Szépl. *siculus* Mon. in Aeolian Islands (Sicily) in 1965]. *Entomophaga*, 11, 411-432.

Nardi, F., Carapelli, A., Dallai, R., Roderick, G.K. & Frati, F. (2005). Population structure and colonization history of the olive fly, *Bactrocera oleae* (Diptera: Tephritidae). *Molecular Ecology*, 14, 2729-2738.

Neuenschwander, P. & Michelakis, S. (1978). Infestation of *Dacus oleae* (Gmel.) (Diptera, Tephritidae) at harvest time and its influence on yield and quality of olive oil in Crete. *Journal of Applied Entomology*, 86, 420-433.

Ortega, M. & Pascual, S. (2014). Spatio-temporal analysis of the relationship between landscape structure and the olive fruit fly *Bactrocera oleae* (Diptera: Tephritidae). *Agricultural and Forest Entomology*, 16, 14-23.

Paredes, D., Cayuela, L., Gurr, G.M. & Campos, M. (2015). Single best species or natural enemy assemblages? A correlational approach to investigating ecosystem function. *BioControl*, 60, 37-45.

Quilici, S. & Rousse, P. (2012). Location of host and host habitat by fruit fly parasitoids. *Insects*, 3, 1220-1235.

Ragusa, S. (1974). Influence of temperature on the oviposition rate and longevity of *Opius concolor siculus* [Hymenoptera: Braconidae]. *Entomophaga*, 19, 61-66.

Raspi, A. & Canale, A. (2000). Effect of superparasitism on *Ceratitis capitata* (Wiedemann) (Diptera Tephritidae) second instar larvae by *Psyttalia concolor* (Szépligeti) (Hymenoptera, Braconidae). *Redia*, 83, 123-131.\

Raspi, A. & Loni, A. (1994). Alcune note sull'allevamento massale di *Opius concolor* Szépligeti (Hym.: Braconidae) e su recenti tentative d'introduzione della specie in Toscana e Liguria [Some notes on the breeding of massale *Opius concolor* Szépligeti (Hym.: Braconidae) and recent tentative introduction of the species in Tuscany and Liguria]. *Frustula Entomologica*, 30, 135-145.

Rugman-Jones, P.F. Wharton, R., van Noort, T. & Stouthamer, R. (2009). Molecular differentiation of the *Psyttalia concolor* (Szépligeti) species complex (Hymenoptera: Braconidae) associated with olive fly, *Bactrocera oleae* (Rossi) (Diptera: Tephritidae), in Africa. *Biological Control*, 49, 17-26.

Russell, M. (2015). A meta-analysis of physiological and behavioral responses of parasitoid wasps to flowers of individual plant species. *Biological Control*, 82, 96-103.

Santos, S.A.P., Pereira, J.A., Torres, L.M. & Nogueira, A.J.A. (2009). Voracity of coccinellid species on different phenological stages of the olive pest *Saissetia oleae* (Homoptera, Coccidae). *Applied Ecology and Environmental Research*, 7, 359-365.

Sime, K.R., Daane, K.M., Messing, R.H & Johnson, M.W. (2006). Comparison of two laboratory cultures of *Psyttalia concolor*. (Hymenoptera: Braconidae), as a parasitoid of the olive fruit fly. *Biological Control*, 39, 248-255.

Sivinski, J. (1991). The influence of host fruit morphology on parasitization rates in the Caribbean fruit fly. *Anastrepha suspensa. Entomophaga*, 36, 447-454.

Sivinski, J., Vulinec, K. & Aluja, M. (2001). Ovipositor length in a guild of parasitoids (Hymenoptera: Braconidae) attacking *Anastrepha* spp. fruit flies (Diptera: Tephritidae) in southern Mexico. *Annals of the Entomological Society of America*, 94, 886-895.

Sivinski, J., Wahl, D., Holler, T., Dobai, S.A. & Sivinski, R. (2011). Conserving natural enemies with flowering plants: Estimating floral attractiveness to parasitic Hymenoptera and attraction's relationship to flower and plant morphology. *Biological Control*, 58, 208-214.

Skouras, P.J., Margaritopoulos, J.T., Seraphides, N.A., Ioannides, I.M., Kakani, E.G., Mathiopoulos, K.D. & Tsitsipis, J.A. (2007). Organophosphate resistance in olive fruit fly, *Bactrocera oleae,* populations in Greece and Cyprus. *Pest Management Science*, 63:42-48.

SPSS Inc., IBM Company (2010). IBM, SPSS Statistic for Windows, version 19.0.0. New York.

Stavraki-Pavlopoulou, H. (1967). Essais préliminaires de lâchers d' *Opius concolor* Szépl. (Hymen.: Braconidae) parasite du *Dacus oleae* Gmel. (Dipt.: Trypedidae) dans l'île de Chalki 1965 [Preliminary tests of releases of *Opius concolor* Szépl. (Hymen.: Braconidae) parasite of *Dacus oleae* Gmel. (Dipt.: Trypedidae) on the island of Halki in 1965]. *Annales de l'Institut Phytopathologique Benaki*, 8, 23-31.

Sterk, G., Hassan. S.A., Baillod, M., Bakker, F., Bigler, F., Blümel, S., Bogenschütz, H., Boller, E., Bromand, B., Brun, J., Calis, J.N.M., Coremans-Pelseneer, J., Duso, C., Garrido, A., Grove, A., Heimbach, U., Hokkanen, H., Jacas, J., Lewis, G., Moreth, L., Polgar, L., Rovesti, L., Samsoe-Petersen, L., Sauphanor, B., Schaub, L., Staübli, A., Tuset, J.J., Vainio, A., Van De Veire, M., Viggiani, G., Viñuela, E. and Vogt, H. (1999). Results of the seventh joint pesticide testing programme carried out by The IOBC/wprs Working Group Pesticides and Beneficials. *BioControl*, 44, 99-117.

Stuhl, C., Sivinski, J., Teal, P. Paranhos, B. & Martin, A. (2011). A compound produced by fruigivorous tephritidae (Diptera) larvae promotes oviposition behavior by the biological control agent *Diachasmimorpha longicaudata* (Hymenoptera: Braconidae). *Environmental Entomology*, 40, 727-736.

Szèpligeti, G. (1910). Sur un Braconide (Hym.) nouveau, parasite du *Dacus oleae* [On a new Braconide (Hym.) Again, parasite *Dacus oleae*]. *Bulletin de la Societe Entomologique de France*, 13, 243-244.

Vayssières, J-F., Adandonon, A., N'Diaye, O., Sinzogan, A., Kooymann, C., Badji, K., Rey, J-Y. & Wharton, R.A. (2012). Native parasitoids associated with fruit flies (Diptera: Tephritidae) in cultivated and wild fruit crops in Casamance, Senegal. *African Entomology*, 20, 308-315.

Viñuela, E., Adán, A., Smagghe, G., González, M., Medina, M.P., Budia, F., Vogt, H. & Del Estal, P. (2000). Laboratory effects of ingestion of azadirachtin by two pests (*Ceratitis capitata* and *Spodoptera exigua*) and three natural enemies (*Chrysoperla carnea*, *Opius concolor* and *Podisus maculiventris*). *Biocontrol Science and Technology*, 10, 165-177.

Viñuela, E., Medina, M.P., Schneider, M., González, M., Budía, F., Adán, A. & Del Estal, P. (2001). Comparison of side-effects of spinosad, tebufenozide and azadirachtin on the predators *Chrysoperla carnea* and *Podisus maculiventris* and the parasitoids *Opius concolor* and *Hyposoter didymator* under laboratory conditions. *IOBC/wprs Bulletin*. 24, 25-34.

Yokoyama, V.Y., Rendon, P.A. & Sivinski, J. (2006). Fruit flies of economic importance: from basic to applied knowledge. *Proceedings of the 7th International Symposium on Fruit Flies of Economic Importance* (pp. 157-164), 10-15 September, Salvador (Brazil).

Yokoyama, V.Y., Rendón, P.A. & Sivinski, J. (2008). *Psyttalia* cf. *concolor* (Hymenoptera: Braconidae) for biological control of olive fruit fly (Diptera: Tephritidae) in California *Environmental Entomology*, 37, 764-773.

Wäckers, F. L. (2005). Suitability of (extra-)floral nectar, pollen, and honeydew as insect food sources. F.L. Wäckers, P.C.J. van Rijn and J. Bruin (Eds.). *Plant-Provided Food for Carnivorous Insects, A Protective Mutualism and its Applications* (pp. 17-74). Cambridge: Cambridge University Press.

Wang, X-G, Levy, K., Son, Y., Johnson, M.W. & Daane, K.M. (2012). Comparison of the thermal performance between a population of the olive fruit fly and its co-adapted parasitoids. *Biological Control*, 60, 247-254.

Wharton, R.A. (2009). Two new species of *Psyttalia* Walker (Hymenoptera, Braconidae, Opiinae) reared from fruit-infesting tephritid (Diptera) hosts in Kenya. *ZooKeys*, 20, 349-377.

Zapata, N., Medina, P., Viñuela, E. & Budia, F. (2005). Toxicidad de malation, pimetrocina, piretrinas naturales + PBO y triflumuron en adultos del parasitoide *Psyttalia concolor* (Szépligeti) (Hym.: Braconidae) según el modo de aplicación [Toxicité des malathion, pymétrozine, pyréthrines + PBO et triflumuron adulte naturel parasitoïde *Psyttalia concolor* (Szépligeti) (Hym. Braconidae) comme mode d'application]. *Boletín de Sanidad Vegetal. Plagas*, 31, 111-118.

Zhang, L. & Raspi, A. (1999). Learning behaviour of *Opius concolor* host discrimination Szlépligeti (Hymenoptera: Braconidae) in host discrimination. *Entomologia Sinica*, 6, 259-266.

BIOGRAPHICAL SKETCHES

Lara Alina de Almeida Pinheiro

CIMO, School of Agriculture, Polytechnic Institute of Bragança

Education: Bachelor in Biological Control, School of Agriculture, Polytechnic Institute of Coimbra

Master in Agroecology, School of Agriculture, Polytechnic Institute of Bragança

Research and Professional Experience: Biological control, Entomology

Professional Appointments: PhD Student

Honors: Higher Grade Master Student (2012)

Publications Last Three Years:

Pinheiro, L. A., Torres, L., Raimundo J. & Santos, S. A.P., 2014. Effects of pollen, sugars and honeydew on lifespan and nutrient levels of *Episyrphus balteatus*. BioControl. DOI: 10.1007/s10526-014-9621-8.

Pinheiro, L. A., Torres, L., Raimundo J. & Santos, S. A.P., 2013. Effect of seven species of the family Asteraceae on longevity and nutrient levels of *Episyrphus balteatus*. BioControl, 58: 797-806. DOI 10.1007/s10526-013-9535-x.

Pinheiro, L. A., Torres, L., Raimundo, J. & Santos, S.A.P., 2013. Effect of floral resources on longevity and nutrient levels of *Episyrphus balteatus* (Diptera: Syrphidae). Biological Control, 67(2): 178-185. DOI 10.1016/j.biocontrol.2013.07.010.

Maria Pilar Medina Vélez

Crop Protection Unit. Department of Agrarian Production.
School of Agricultural, Food and Biosystems Engineering.
Technical University of Madrid (UPM).

Education: Agricultural Engineering. Phd in Applied Entomology.

Address: Crop Protection Unit. Department of Agrarian Production. School of Agricultural, Food and Biosystems Engineering. Ciudad Universitaria, s/n. 28040. Madrid. Spain.

Research and Professional Experience: Associate Professor at UPM.

Professional Appointments: Secretary of the Departament of Agrarian Production.

Honors: PhD extraordinary award

Publications Last Three Years:

1. Maia, J.B., Carvalho, G.A., Medina, P., Garzón, A., Contijo, P.C., Viñuela E. 2016. Lethal and sublethal effects on *Chrsyoperla carnea*

larva and the influence of rainfastness in the pesticide degradation pattern over time. *Ecotoxicology* DOI 10.1007/s10646-016-1641-y.

2. Garzon, A., Budia, F. Morales, I., Fereres, A., Viñuela, E., Medina, P. 2016. Do *Chrysoperla carnea* and *Adalia bipunctata* influence the spread of Cucurbit aphid-borne yellows virus and its vector *Aphis gossypii*?" Annals of Applied Biology DOI 10.1111/aab.12284.
3. Wanumen, A.C., Carvalho, G.A., Medina, P., Viñuela, E., Adan, A. 2016. Residual acute toxicity of some modern insecticides toward two mirid predators of tomato pests. Journal of Economic Entomology. DOI: 10.1093/j33/tow059.
4. Garzón,A., Freire, B., Carvalho, Ga., De Oliveira, Rl., Medina, P., Budia, F. 2015. Development and reproduction of *Chrysoperla externa* (Hagen) (Neuroptera: Chrysopidae) fed on *Myzus persicae* (Sulzer) (Hemiptera: Aphididae) vectoring Potato leafroll virus (PLRV). Neotropical Entomology. DOI: 10.1007/s13744-015-0329-y.
5. Fernadéz, M., Medina, P., Fereres, F., Smagghe, G., Viñuela, E. 2015. Are mummies and adults of *Eretmocerus mundus* (Hymenoptera: Aphelinidae) compatible with insecticides? Journal of Economic Entomology. DOI: 10.1093/jee/tov181.
6. Garzón, A., Medina, P., Amor, F., Viñuela, E., Budia, F. 2015. Toxicity and sublethal effects of six insecticides to last instar larvae and adults of the biocontrol agents *Chrysoperla carnea* (Stephens) (Neuroptera: Chrysopidae) and *Adalia bipunctata* (L.) (Coleoptera: Coccinellidae). Chemosphere 132: 87-93. DOI: 10.1016/j. chemosphere.2015.03.016.
7. Garzón, A., Budia, F., Medina, P., Morales, I., Fereres, A., Viñuela, E. 2014. The effect of *Chrysoperla carnea* (Neuroptera: Chrysopidae) and *Adalia bipunctata* (Coleoptera: Coccinellidae) on the spread of cucumber mosaic virus (CMV) by *Aphis gossypii* (Hemiptera: Aphididae. DOI: 10.1017/s0007485314000534.
8. Bengochea, P., Sanchez-Ramos, I., Saelices, R., Amor, F., del Estal, P., Viñuela, E., Adán, A., Budia, F., Medina, P. 2014. Is Emamectin Benzoate Effective Against the Different Stages of *Spodoptera exigua* (Hübner) (Lepidoptera, Noctuidae)? Irish Journal of Agriculture and Food Research 53: 37–49.
9. Olombrada, M., Martinez Del Pozo, A., Medina, P., Budia, F., Gavilanes, J.G., García-Ortega, L. 2014. Fungal ribotoxins: Natural protein-based weapons against insects (mini review). Toxicon 83: 69-74. DOI: 10.1016/j.toxicon.2014.02.022.

10. Bengochea, P., Saelices, R., Amor, F., Adan, A., Budia, F., Del Estal, P. Viñuela, E., Medina, P. 2014. Non-target effects of kaolin and coppers applied on olive trees for the predatory lacewing *Chrysoperla carnea*. Biocontrol Science and Technology 24 (6): 625-640. DOI: 10.1018/09583157.2014.884212.
11. Bengochea, P., Budia, F., Viñuela, E., Medina, P. 2014. Are kaolin and copper treatments safe to the olive fruit fly parasitoid *Psyttalia concolor*? Journal of Pest Science 87 (2): 351-359. DOI: 10.1007/s10340-013-0543-5.

In: Natural Enemies
Editor: Sónia A.P. Santos

ISBN: 978-1-63485-921-9

Chapter 6

BIOLOGICAL CONTROL BASED ON NATIVE PARASITOIDS ASSOCIATED WITH *DRYOCOSMUS KURIPHILUS* YASUMATSU IN INVADED COUNTRIES

Ana Santos[1], José Alberto Pereira[2], Sónia A. P. Santos[1], Ambra Quacchia[3] and Albino Bento[1,*]

[1]Mountain Research Center (CIMO), School of Agriculture, Polytechnic Institute of Bragança, Bragança, Portugal

[2]School of Agriculture, Polytechnic Institute of Bragança, Bragança, Portugal

[3]GreenWood Service Srl., Italy

ABSTRACT

The chestnut gall wasp, *Dryocosmus kuriphilus* Yasumatsu (Hymenoptera: Cynipidae), is a gall-inducing insect that attacks buds of trees from the genus *Castanea*, and can significantly reduce growth, fruiting and consequently the production of chestnut trees. *D. kuriphilus* is native to China and its dispersion started in 1941, in Japan, followed by South Korea (1958), the US (1974) and Italy (2002). In 2005, this pest

* Corresponding author: Albino Bento. Mountain Research Center (CIMO), School of Agriculture, Polytechnic Institute of Bragança, Campus de Santa Apolónia, 5300-252 Bragança, Portugal. Email: bento@ipb.pt.

was detected in France from where it spread to other countries, including Spain and Portugal. This insect is part of the A2 list of the European and Mediterranean Plant Protection Organization (EPPO) and is considered one of the most dangerous organisms to the genus *Castanea*. Its life cycle has been studied in several countries and reveals a common pattern: there is only one generation per year, during which females emerge from the galls from late May until the beginning of August. In the following spring, when the chestnut begins to burst, the wintering larvae induce the formation of galls. Biological control of this pest is essentially based on releases of *Torymus sinensis* Kamijo (Hymenoptera: Torymidae), a parasitoid from China that has been introduced in countries where the pest was already present. This parasitoid is reported to be highly specific and its life cycle is synchronized with that of *D. kuriphilus*. However, studies performed in several regions showed that there is a very diverse assemblage of native parasitoids associated with *D. kuriphilus*. Most of these parasitoids are also associated with other tree galls such as the oak galls and can act as biological control agents of *D. kuriphilus*. The most frequent species belong to the families Torymidae, Pteromalidae, Ormyridae, Eupelmidae and Eurytomidae. The aim of this chapter is to study the communities of native parasitoid associated with *D. kuriphilus* in invaded countries.

Keywords: chestnut, *Castanea sativa*, Torymidae, *Torymus sinensis*, gall, *Eupelmus urozonus*

1. INTRODUCTION

The chestnut *Castanea* spp. Mill. (Family Fagaceae) has a great socio-economic, environmental and landscape importance in many areas of the world, playing a multipurpose role of fruit and wood production, agro-forestry co-production (e.g., edible mushrooms or honey) and spare time activities (Bounous, 2005; Monteiro & Patrício, 2007).

There are 13 different species of the genus *Castanea* distributed throughout the world, but the most important ones are: *C. crenata* Siebold and Zucc. (the Japanese chestnut, native to Korea and Japan), *C. mollissima* Blume (the Chinese chestnut, native to China, Taiwan and Korea), *C. dentata* (Marsh.) Borkh. (the American chestnut, native to North America) and *C. sativa* Mill. (the European chestnut, native to Europe) (Cortizo et al., 1999).

Since the Middle Ages, the chestnut fruit has been an important food resource in European rural areas, especially in mountain regions, during winter

(Monteiro & Patrício, 2007). Due to its nutritional quality (with high levels of starch, fiber and minerals) and benefits for human health (presence of vitamins and ω-3 and ω-6 polyunsaturated fatty acids), the consumption of the chestnut fruit has increased in recent years (Borges et al., 2008).

Despite their importance, chestnut trees are being threatened by several pests and diseases such as the chestnut blight, caused by the pathogenic fungus *Cryphonectria parasitica* (Murr.) and the *Phytophthora* dieback, caused by *Phytophthora cinnamomi* Rands and pests such as the chestnut tortrix *Cydia splendana* (Hübner), the beech moth *Cydia fagiglandana* (Zeller), the chestnut leafroller *Pammene fasciana* (L.) (Lepidoptera: Tortricidae), the chestnut weevil *Curculio elephas* (Gyllenhal) (Coleoptera: Curculionidae) (Mansilla et al., 2000) and the Asian chestnut gall wasp *Dryocosmus kuriphilus* Yasumatsu (Hymenoptera, Cynipidae) (Yasumatsu, 1951).

Among pests, *D. kuriphilus*, native to China, is a key pest of the chestnut (EPPO, 2005). This insect is on the European and Mediterranean Plant Protection Organization (EPPO) A2 list of pests recommended for regulation as a quarantine pest and is considered one of the most harmful organisms to the genus *Castanea* (EPPO, 2014). *D. kuriphilus* disrupts the growth of the chestnut by inducing gall formation on new shoots, suppressing nut production that, according EPPO (2005) can reach 70%, and causing a gradual decline in its vigour and growth (EFSA, 2010). This species was established in the mid-twentieth century in Japan (1941), Korea (1958) and in the US (1974) (Aebi et al., 2006). In Europe, it was first reported in Italy in 2002 (Brussino et al., 2002) and then in France in 2005 (Aebi et al., 2006). In 2010, the European Food Safety Authority (EFSA) Panel on Plant Health delivered a scientific opinion on the risk posed by *D. kuriphilus* to the European Union (EU) whose aim was to hinder the progression of *D. kuriphilus* throughout other European countries. However, the invasion has proceeded in several regions and the pest is now present in most of the European countries (Croatia, Czech Republic, France, Germany, Hungary, Italy, Netherlands, Portugal, Slovenia, Spain, Switzerland and United Kingdom).

Biological control of this pest is essentially based on releases of *Torymus sinensis* Kamijo (Hymenoptera: Torymidae), a parasitoid native to China that has been introduced in countries where the pest was already present. This parasitoid is reported to be highly specific and its life cycle is synchronized with that of *D. kuriphilus* (EPPO, 2005). Biological control is considered as a potential means of reducing the levels of infestation caused by *D. kuriphilus* in invaded areas. Classical biological control programs, based on the importation of the parasitoid *T. sinensis* from its native areas, have been carried out in different countries such as Japan (1975), the US (in the late 1970s), Italy (2005), France (2010), Croatia (2014), Hungary (2014), Portugal (2015) and Spain (2015).

In the scope of biological control programs, a different strategy based on managing of *D. kuriphilus* through its native parasitoids occurring in each invaded region is under investigation. In China, *D. kuriphilus* is attacked by 15 parasitoid species, belonging to five Chalcidoidea families (i.e., Torymidae, Ormyridae, Eurytomidae, Eulophidae and Eupelmidae) (Murakani et al., 1980). Native parasitoids that normally attack galls induced by related cynipid gall wasps on oaks (Fagaceae, *Quercus* spp.) have parasitized *D. kuriphilus* galls around the world (Aebi et al., 2006). The aim of this chapter is to study the community of native parasitoids associated with *D. kuriphilus* in affected countries.

2. *Dryocosmus kuriphilus*: An Overview of the Bioecology of the Pest

D. kuriphilus (Figure 1) is a univoltine insect, with parthenogenetic reproduction. Adult females live for 2-10 days (Yasumatsu, 1951). Adults emerge in early summer (June and July) and lay eggs in chestnut buds that will develop in the following spring. Each female may lay more than 100 eggs which are oval, milky white, between 0.1-0.2 mm long, with a long stalk and hatch in 30-40 days (Ôtake, 1980; Ôtake, 1989 EPPO, 2005). The first-instar larva is subglobular, measuring 0.2-0.6 mm in length, and develops slowly from the summer (July-August) to the late March-April of the next year (Viggiani, 2010). The terminal larva is 2.5 mm long when fully grown, milky white, without eyes and legs. At bud burst, in spring, larval feeding induces the formation of galls with around 5-20 mm in diameter (EFSA, 2010). Pupation occurs within the gall from mid-May to mid-July, and adult females (2.5-3 mm long) emerge and fly from the end of May until the end of July, depending on locality and chestnut cultivar (EPPO, 2005; EFSA, 2010) (Figure 2).

D. kuriphilus can disperse through active flight, with an average rate of dispersion estimated between 8-25 km/year, although its geographic range is mainly due to the movement of infested plants caused by human activity. Nursery plants need to be cultivated using an anti-insect netting to ensure the isolation of crops or in greenhouses with their ventilating areas covered with anti-insect netting (EFSA, 2010; Perez & Lopez, 2015). Introduction of plants from Asia and America into Europe is effectively prohibited (EPPO, 2005).

The main symptom caused by this pest is the development of galls in branches, petioles or the central veins of leaves, from mid-April till the end of June (Figure 3). Galls are hypertrophic tissues and can measure between 5 and 20 mm in diameter. They can be green or reddish, or they can be green and then turn reddish. From autumn, galls dry up and can remain in the tree for several years (Brussino et al., 2002). No symptoms are detected from the egg-laying to the egg hatching (Quacchia et al., 2011; Pérez & López, 2015).

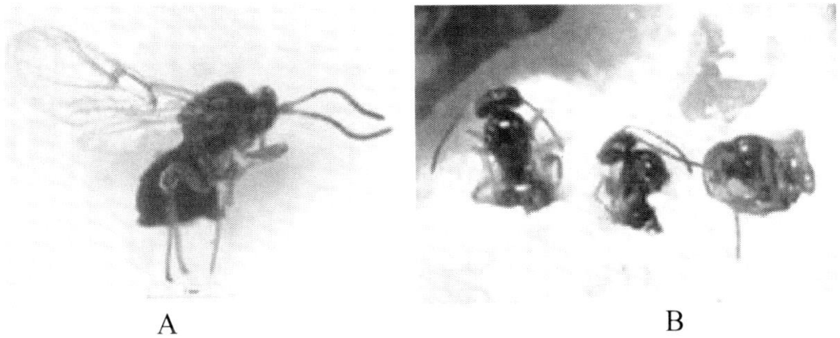

Figure 1. *Dryocosmus kuriphilus* - general aspect (A) and *D. kuriphilus* emerging from chestnut galls collected in Entre-Douro-e-Minho, Portugal, 2014 (B). Photos by A. Santos and A. Dinis.

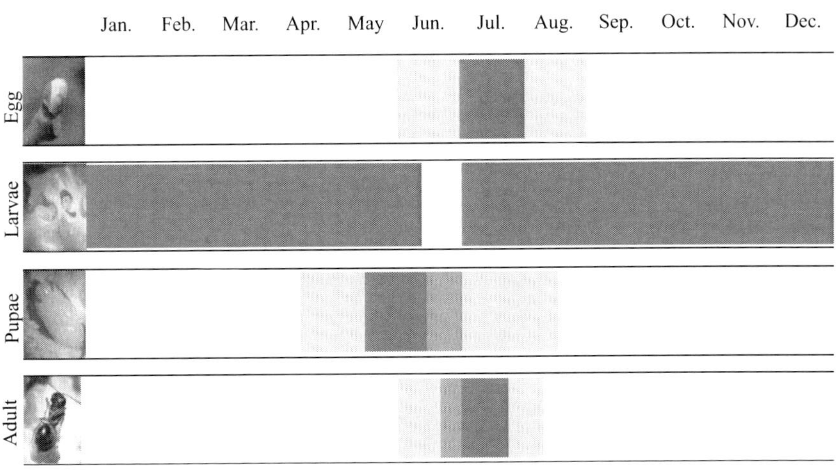

Figure 2. *Dryocosmus kuriphilus* life cycle observed in the northwest of Portugal (Entre-Douro-e-Minho region), 2014/2015. Photos by A. Dinis.

Figure 3. *Dryocosmus kuriphilus* galls affecting leaves of the chestnut tree. Photo by A. Bento.

The chestnut gall proliferation can cause (Gehring et al., 2014; Graziosi et al., 2011; Pérez & López, 2015):

- Reduction of photosynthetic area, leading to a decrease of carbohydrates available for the growth of trees or fruits;
- Decrease of the number of hibernating buds, which results in a reduction of the number of new buds the following year;
- Gradual reduction of the growing season of trees;
- Reduction of the number of inflorescences and, as a result, fewer fruits;
- Increase susceptibility to fungal cankers;
- Modification of the architecture of trees.

The susceptibility of the chestnut to the attack of *D. kuriphilus* can vary with the cultivar. Santor et al. (2009) evaluated the susceptibility of 41 chestnut cultivars from different regions of Italy and other European countries as well as hybrid cultivars. Two cultivars (Marsol and Madonna) were found to be very susceptible to the attack of *D. kuriphilus*, followed by Marigoule and Marrone di Zocca cultivars. On the other hand, Bouche de Bétizac cultivar was not infested by *D. kuriphilus* during the four years of observations.

3. NATURAL ENEMIES OF *DRYOCOSMUS KURIPHILUS*

Surveys of native communities of hymenopteran parasitoids that emerged from *D. kuriphilus* galls were preformed in several countries and different species belonging to the superfamily Chalcidoidea were identified. Chalcidoidea represents a significant portion of the biological diversity found in terrestrial ecosystems and it plays a major role in regulating populations of terrestrial insects. Chalcidoids are known to attack a broad spectrum of hosts mainly belonging to the following orders of Insecta: Coleoptera, Diptera, Hemiptera, Hymenoptera, Lepidoptera, Neuroptera, Odonata, Orthoptera, Psocoptera, Siphonaptera, Strepsiptera and Thysanoptera, as well as egg sacs of spiders (Araneae), ticks and gall-forming mites (Acari), cocoons of pseudoscorpions (Pseudoscorpiones), and gall-forming Anguinidae (Nematoda). Moreover, some species are phytophagous, either as gall-formers or as inquilines in galls (Gibson et al., 1997). This superfamily includes solitary and gregarious species, ectoparasitoides and endoparasitoids, primary, secondary and tertiary parasitoids, polyembryonic species, species with planidial larvae (which is a quite mobile first instar parasitoid larvae) and specialist, generalist or opportunist species (Gibson et al., 1997). It also includes parasitoid species that are considered successful biological control agents causing high mortality rate in cynipid gall hosts and can be found in a wide geographical range (Greathead, 1986; Stone et al., 2002).

Native parasitoids need time to recognize and adapt to new potencial hosts, and this process is often expressed differently in diverse species and also depends on human environmental modifications (Folcher et al., 2011). Most of the parasitoid species found in cynipid galls on oaks do not attack other hosts on other plants. However, a small number of species, such as *Eupelmus urozonus* Dalman and *Eupelmus vesicularis* (Retzius) (Eupelmidae), are exceptionally polyphagous and attack a range of endophytic hosts (Askew, 1984).

To date, Chalcidoidea species attacking *D. kuriphilus* belong to the Eulophidae, Eupelmidae, Eurytomidae, Ormyridae, Pteromalidae and Torymidae families (Gómez et al., 2006) (Figure 4).

The Eulophidae family is dominated by parasitoids of Lepidoptera, Coleoptera, Diptera and Hymenoptera, however, many other hosts are parasitized. A large number of Eulophidae species are parasitoids of hosts covered by plant tissues, such as leaf miners, wood borers, leaf rollers and gall-inducers. This family also contains phytophagous species, including gall-inducers, seed predators, and inquilines (Gibson et al., 1997).

The Eupelmidae family parasitizes Lepidoptera, Coleoptera, Hymenoptera and other orders of insects and spiders (Gómez et al., 2006). They are ectoparasitoids of larvae or pre-pupae of insects usually covered by plant tissue or cocoons, or are endoparasitoids or predators of insects or spider eggs.

Many species are polyphagous, apparently parasitizing diverse hosts of similar morphological appearance in a particular ecological niche (Gibson et al., 1997).

The Eurytomidae family includes phytophagous, gall-inducers, granivorous, ectoparasitoids, hyperparasitoids of gall-inducer larvae, leaf miners or wood borers (Gómez et al., 2006). Most species of this family are solitary parasitoids and even though more than one egg may be placed on the host, larval competition ensures that only a single parasitoid develops (Gauld & Bolton, 1988).

The Ormyridae family parasitizes insects in plant galls or in seeds. Most Ormyridae are associated with galls of Cynipidae, with the majority of the species being restricted to cynipid galls on oaks and others on the Rosaceae family or blueberry plant (Gibson et al., 1997).

The Pteromalidae family has a great diversity of hosts, including many different orders of insects and even egg sacs of spiders. This family parasitizes larvae or pupae of small insects feeding in or on plants (Gibson et al., 1997). There are some solitary species, others are gregarious, ectoparasitoids, endoparasitoids, primary and secondary parasitoids, and even predators. Females can lay about 700 eggs outside or inside the host (Gauld & Bolton, 1988).

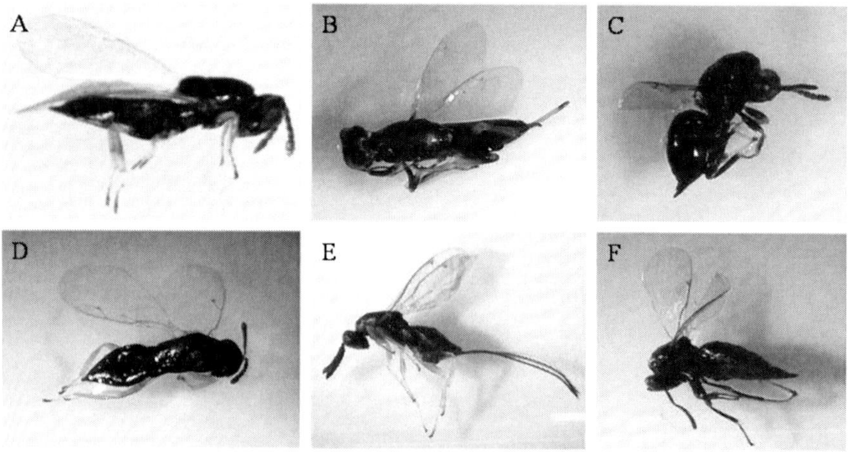

Figure 4. Specimens of natural enemies of *Dryocosmus kuriphilus* belonging to the Chalcidoidea superfamily: A - Eulophidae; B - Eupelmidae; C - Eurytomidae; D - Pteromalidae; E - Torymidae; F - Ormyridae. Photos by A. Santos.

The Torymidae family includes both entomophagous and phytophagous species. The entomophagous species are either solitary or gregarious ectoparasitic idiobionts. Many species were reared from gall-inducer insects including Cynipidae, Tanaostigmatidae, Eurytomidae, Cecidomyiidae, Tephritidae, and Psyllidae. Other species were reared from insect larvae and pupae in protected places such as seed pods, plant stems and solitary wasp and bee nests (Gibson et al., 1997).

3.1. Natural Enemies of *Dryocosmus kuriphilus* in China

In its native country, *D. kuriphilus* is distributed throughout 16 provinces covering all the productive areas, where it has caused production losses between 15% and 30% every year (Zhi-Yong, 2009). From 1975 to 1979, nine parasitoid species were identified from galls collected in surveys performed in several Chinese provinces, namely: *E. urozonus* (Eupelmidae), *Eurytoma brunniventris* Ratzeburg, *Eurytoma setigera* Mayr, *Sychophila varieata* (Curtis) (Eurytomidae), *Megastigmus maculipennis* Yasumatsu and Kamijo, *Megastigmus nipponicus* Kamijo, *Torymus geranii* (Walker), *Torymus* (*Syntomaspis*) sp. (Torymidae), *Ormyrus punctiger* (Westwood) (Ormyridae), *Tetrastichus* sp. (Eupelmidae) (Murakani et al., 1980). In 1979, the parasitism rate caused by *Torymus* (*Syntomaspis*) sp. varied from 4.2% to 52.1%, depending on the province, and was indicated as a promising natural enemy of *D. kuriphilus*. This species was later indentified as *T. sinensis* (Murakami, 1981). In Habei Province, the parasitism rate varied from 24.2% in 1978 to 81.2% in 1980, which was accompanied by a decrease of the ratio of damadged shoots from 56.8% to 9.9% respectively (Zhi-Yong, 2009).

3.2. Introduction and Efficiency of *Torymus sinensis*

T. sinensis (Figure 5) is considered an efficient biological control agent of *D. kuriphilus*, reducing its population to economically sustainable levels (Murakami et al., 2001; Gibbs et al., 2011). Among the parasitic assemblage of natural enemies of *D. kuriphilus*, *T. sinensis* was the dominant species in several areas of China (Zhi-Yong, 2009).

Figure 5. *Torymus sinensis* female on a chestnut gall. Photo by A. Dinis.

The introduction of *T. sinensis* has already occurred in several countries as a means to control *D. kuriphilus* (Matošević et al., 2014). In Japan, a preliminary introduction occurred in 1975 and then in 1979 and 1981. In the following years, the population of *T. sinensis* increased and, by 1989, it dominated the parasitoid community of *D. kuriphilus* at the release sites and expanded its geographical area 12 km apart from the release site (Moriya et al., 2003). Furthermore, the infestation rate decreased to less than 1% in 1992, which was mainly attributed to the parasitoid (Moriya et al., 2003).

In the late 1970s, *T. sinensis* was introduced to Georgia, in the US, from Japan and in 2005 was confirmed as the prevalent parasitoid species of *D. kuriphilus* in galls collected in Virginia and Ohio (Cooper and Rieske, 2007a).

In Europe, *T. sinensis* that emerged from galls sent from Japan was first introduced in the province of Cuneo, in Italy, in 2005. The parasitism rate of *T. sinensis* increased in values from 75% to 98.6% in 5-7 years and drastically reduced the infestation level of *D. kuriphilus* (Quacchia et al., 2008; Quacchia et al., 2014a). France has implemented classical biological control using *T. sinensis* since 2010. In the following two years, 42 releases were performed in different orchards and the first results indicated that the population of *T. sinensis* has remained in 50% of the release sites (Borowiec et al., 2014). In Croatia and Hungary, *T. sinensis* was introduced in 2014 and for the next few years, research will proceed in order to study the parasitism rate of parasitoid species (Matošević et al., 2014).

These introductions have shown that *T. sinensis* can successfully control *D. kuriphilus* mainly because it is phenologically well synchronised with its host (with one emergence period) and is able to disperse rapidly throughout the invaded territory (Matošević et al., 2014). *T. sinensis* is univoltine (i.e., only one generation per year) like its host. After emergence, that takes place in early spring, and mating, females lay eggs inside the larval chamber of newly formed galls, usually one egg per host larva. After hatching, the larva feeds ectoparasitically on the host larva and pupates in the host larval chamber during winter (Ferracini et al., 2015). According to recent observations, the pupation can occur since autumn. The mean life span of *T. sinensis* in open field is about one month, while in the laboratory they can live for over 4 months when feed on honey and kept at constant temperature of 15°C (Matošević et al., 2014). Females locate the host with a combination of visual and olfactory stimuli emitted from fresh galls and chestnut foliage (Matošević et al., 2014). When mating does not occur, females lay unfertilized eggs, parthenogenetically producing only males. This particularity should be considered when planning rearing and releases into new sites in order to obtain a proper male/female ratio (Matošević et al., 2014).

Furthermore, *T. sinensis* presents a prolonged diapause, mainly as late instar larva. Diapause can be extended for 12 months and adults emerge in spring as usual, presenting a two-year life cycle. Prolonged diapause is commonly seen as strategy which provides demographic refuges for insect populations living in variable environments. Among oak gall parasitoids it is an extremely rare strategy, presumably because most of these parasitoids are polyphagous and able to use hosts in a range of gall types. *T. sinensis*, in contrast, is a parasitoid of apparently only one host species and if the availability of *D. kuriphilus* galls is unpredictable, extended diapause might be expected to have selective benefits (Quacchia et al., 2014). Individuals that emerged in the second year were able to mate, presenting the same parasitism behavior as those which emerged during the first year (Ferracini et al., 2015).

Although the ability to adapt to new environments and the efficacy of *T. sinensis* as biological control agent has been demonstrated, the risks associated with its introduction should be taken into account. Risks include host-shifting to non-target species and hybridization with native *Torymus* species. Some studies, such as that performed by Yara et al. (2010), proved the hybridization in Japan with native *Torymus beneficus* Yasumatsu. In Europe, hybridization between *T. sinensis* and *T. cyaneus* was documented in one case (Aebi et al., 2013) however this result has not been confirmed. A recent study recorded the emergence of three individuals of *T. sinensis* from galls of *Biorhiza pallida* L. (Hymenoptera: Cynipidae) in 2013. The survey was repeated in 2014 and did not reveal any further emergence (Ferracini et al., 2015); therefore, these three oviposition events may thus be on a less preferred host and are likely to be only very occasional (Everatt, 2015).

Host-specificity is continuum from extreme specialists with a host-range restricted to a single host species, to so-called generalists which have a broad host-range. However, the host-specificity of an insect can be further differentiated according to how acceptable and/or suitable hosts are relative to each other. For example, an insect which performs equally well on all host species is less host-specific than an insect for which only one of many species is a good host, even though their host ranges are identical (Van Klinken, 2000).

Mass rearing: Withered galls are collected during winter in highly parasitized orchards or woods, cleaned to remove contaminants (e.g., unwanted predators) and stored in cardboard boxes provided with removable skylights. These cardboard boxes are placed under environmental conditions (usually in a sheltered location) or stored in a climatic chamber in order to control (by increasing/decreasing the temperature) the emergence of the parasitoids. When adults emerge, they are attracted to the light and move into the skylights. The skylights are then collected and replaced with new ones. In the laboratory, the adults are coupled, at a ratio of about 2:1 females to males, into plastic tubes. Adults are fed by a feeder soaked of honey placed in the cap of the tubes. Storing take place in a climatic chamber (10-15°C) until release. The field release, should match the very beginning of galls formation. Each release usually has a ratio of 100 females to 50 males.

3.3. Natural Enemies of *Dryocosmus kuriphilus* in Invaded Areas

Fifty seven parasitoid species were identified from chestnut galls collected in several countries all around the world (Annex 1). In China, 15 species were collected from chestnut galls, 21 in Japan and 12 in Korea (Schonrogge et al., 2006; Matsuo et al., 2011). In this part of Asia, 10 species, namely: *Eurytoma schaeferi* Yasumatsu and Kamijo, *Eurytoma setigera* Mayr, *Sycophila concinna* (Boheman), *Arthrolytus usubai* Kamijo, *Caenacis peroni* Kamijo, *Mesopolobus yasumatsui* Kamijo, *Torymus beneficus* Yasumatsu and Kamijo, *Torymus koreanus* Kamijo, *Megastigmus maculipennis* Yasumatsu and Kamijo and *Megastigmus nipponicus* Yasumatsu and Kamijo were exclusive since they were only found in those countries. However, some palaearctic species of parasitoids that are effective parasitoids of *D. kuriphilus* in Asia are also very common and widespread in Europe. For instance, throughout Europe, *E. urozonus* is one of the most common *D. kuriphilus* native parasitoids and this species very often also acts as a facultative hyperparasitoid

CONCLUSION

The chestnut gall wasp, *D. kuriphilus* is a dangerous pest that causes severe damage to the production of fruits all over the world. Biological control of this pest has been essentially based on releases of *T. sinensis*, a parasitoid native to China that has been introduced in countries where the pest had already installed. In each invaded region, several studies have shown that *D. kuriphilus* also has other natural enemies which contribute to the reduction of its population; *E. urozonus* is one of the most common species of native parasitoids found in several European and Asian countries. In the northwest of Portugal, chestnut galls presented a high parasitism rate comprising of seven parasitoid species. The study of native parasitoids of *D. kuriphilus* is an important issue for a sustainable management of chestnut orchards and more studies are needed in order to clarify the interactions established between the chestnut, the pest and its parasitoids.

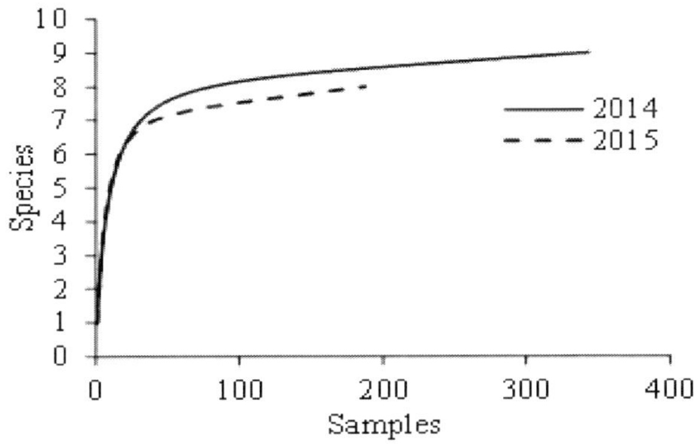

Figure 6. Species accumulation curves of parasitoid species collected from chestnut galls in the northwest of Portugal (Entre-Douro-e-Minho region), in 2014 and 2015.

Annex 1. Parasitoid species identified in several countries from Asia, Europe and the US

Family/Species	China	Japan	Korea	The US	Italy	Slovenia	Hungary	Portugal	Croatia	References
Eulophidae										
Aprostocetus sp.					X					[6, 10]
Aprostocetus biorrhizae						X				[13]

Annex 1. (Continued)

Family/Species	China	Japan	Korea	The US	Italy	Slovenia	Hungary	Portugal	Croatia	References
Aprostocetus aethiops						X				[13]
Aprostocetus glandicola						X				[13]
Aulogymnus arsames					X					[6, 10]
Aulogymnus skianeuros						X				[13]
Aulogymnus sp.					X					[6, 10]
Baryscapus pallidae					X					[3, 4]
Baryscapus sp.					X	X				[6, 10, 13]
Pediobius chilaspidis					X					[6, 10]
Pediobius saulius					X	X				[6, 10, 13]
Pediobius sp.					X					[6, 10]
Pnigalio minio				X						[12]
Tetrastichus sp.	X	X	X							[1, 2]
Eupelmidae										
Eupelmus annulatus					X	X	X	X	X	[6, 8, 10, 13, 14, 15, 16, 17]
Eupelmus spongipartus					X					[6, 10]
Eupelmus splendens					X	X				[6, 10, 13]
Eupelmus urozonus	X	X	X		X	X	X	X	X	[1, 2, 3, 4, 6, 8, 10, 11, 13, 14, 15, 16, 17]
Eupelmus flavipes					X					[6, 10]
Eupelmus rostratus					X					[11]
Eupelmus sp.	X	X	X	X						[1, 2, 7, 9, 12]
Eupelmus (=Macroneura) vesicularis						X				[13, 14]
Eurytomidae										
Eurytoma brunniventris	X	X	X		X		X	X	X	[1, 2, 3, 4, 6, 7, 8, 10, 11, 15, 16, 17]

Family/Species	China	Japan	Korea	The US	Italy	Slovenia	Hungary	Portugal	Croatia	References
Eurytoma adleriae					X					[6, 10]
Eurytoma pistacina					X				X	[3, 4, 6, 10, 17]
Eurytoma schaeferi		X								[1, 2]
Eurytoma setigera	X	X	X							[1, 2]
Eurytoma sp.	X							X		[9, 16]
Sycophila variegata	X				X				X	[1, 2, 3, 4, 6, 10, 11, 17]
Sycophila biguttata					X				X	[3, 4, 6, 10, 11, 14, 17]
Sycophila iracemae					X			X		[3, 4, 7, 10, 16]
Sycophila mellea				X						[12]
Sycophila flavicollis					X	X				[13, 14]
Sycophila binotata					X					[14]
Sycophila concinna		X								[1, 2]
Ormyridae										
Ormyrus nitidulus					X					[6, 10, 14]
Ormyrus pomaceus	X	X	X		X	X	X	X	X	[1, 2, 3, 4, 6, 7, 8, 10, 11, 13, 14, 16, 17]
Ormyrus labotus				X						[12]
Ormyrus sp.					X					[15]
Pteromalidae										
Amblymerus sp.		X								[1, 2]
Arthrolytus usubai		X								[1, 2]
Caenacis peroni		X	X							[1, 2]
Cecidostiba fushica		X								[1, 2]
Cecidostiba semifascia		X				X				[1, 2, 13]
Cecidostiba sp.					X					[6, 10]
Mesopolobus amaenus					X				X	[6, 7, 10, 11, 14, 17]
Mesopolobus dubius									X	[17]

Annex 1. (Continued)

Family/Species	China	Japan	Korea	The US	Italy	Slovenia	Hungary	Portugal	Croatia	References
Mesopolobus mediterraneus					X					[3, 4, 10]
Mesopolobus yasumatsui		X								[1, 2]
Mesopolobus sericeus					X	X			X	[3, 4, 6, 10, 11, 12, 14, 17]
Mesopolobus tarsatus					X					[3, 4, 6, 7, 10, 14, 15]
Mesopolobus albitarsis						X				[13]
Mesopolobus fasciiventris						X				[13, 14, 15]
Mesopolobus tibialis					X		X		X	[6, 8, 10, 14, 15, 17]
Mesopolobus sp.								X		[16]
Pteromalus sp.				X						[12]
Torymidae										
Torymus auratus				X	X				X	[3, 4, 10, 11, 12, 14, 15, 17]
Torymus flavipes				X	X		X	X	X	[3, 4, 5, 6, 7, 8, 10, 11, 12, 15, 16, 17]
Torymus scutellaris					X					[3, 4, 10]
Torymus eucarum					X					[4]
Torymus geranii		X		X	X				X	[1, 2, 5, 6, 7, 12, 17]
Torymus beneficus	X	X								[1, 2]
Torymus sinensis	X	X	X	X						[1, 2, 4, 12, 13]
Torymus koreanus	X	X	X							[1, 2]
Torymus sp.	X	X	X		X					[1, 2, 7, 9]
Torymus formosus				X						[12]
Megastigmus dorsalis					X		X	X	X	[4, 6, 7, 8, 10, 11, 15, 16, 17]
Megastigmus maculipennis	X	X	X							[1, 2]

Family/Species	China	Japan	Korea	The US	Italy	Slovenia	Hungary	Portugal	Croatia	References
Megastigmus sp.	X									[9]
Megastigmus nipponicus	X	X	X							[1, 2]
Ichneumonidae										
Orthopelma mediator					X					[11]

References: 1 - Aebi et al. (2006); 2 - Matsuo et al. (2011); 3 - Aebi et al. (2007); 4 - Speranza et al. (2009); 5 - Santi and Maini (2012); 6 - Quacchia et al. (2013); 7 - Francati et al. (2015); 8 - Bujdosó et al. (2013); 9 - Chang-Ti et al. (2013); 10 - Quacchia et al. (2014b); 11 - Boriani et al. (2013); 12 - Cooper and Rieske (2007a); 13 - Kos et al. (2015); 14 - Palmeri et al. (2014); 15 - Panzavolta et al. (2013); 16 - Santos (2015); 17 - Matošević and Melika (2013)

REFERENCES

Aebi, A., Schonrogge, K., Melika, G., Alma, A., Bosio, G., Quacchia, A., Picciau, L., Abe, Y., Moriya, S., Yara, K., Seljak, G. & Stone, G. N. (2006). Parasitoid recruitment to the globally invasive chestnut gall wasp *Dryocosmus kuriphilus*. In: K. Ozaki, J.Yukawa, T. Ohgushi, and P.W. Price (Eds.) *Galling Arthropods and Their Associates - Ecology and Evolution* (pp. 103-121). Tokyo: Springer-Verlag.

Aebi, A., Schonrogge, K., Melika, G., Quacchia, A., Alma, A. & Stone, G.N. (2007). Native and introduced parasitoids attacking the invasive chestnut gall wasp *Dryocosmus kuriphilus*. *Bulletin OEPP/EPPO*, 37, 166-171.

Aebi, A., Schönenberger, N. & Bigler, F. (2013). Towards an environmental risk assessment of *Torymus sinensis* against the chestnut gall wasp *Dryocosmus kuriphilus* in Switzerland. Report for the FOEN, 34 p.

Askew, R.R (1961). On the biology of the inhabitants of oak galls of Cynipidae (Hymenoptera) in Britain. *Transactions of the Society for British Entomology*, 14, 237-268.

Askew, R.R. (1984). The biology of gall wasps. In: E.T.N. Anantakrishnan (Ed.) *Biology of Gall Insects*. pp. 223-271. London: Edward Arnold.

Borges, O., Gonçalves, B., Carvalho, J.L.S., Correia, P. & Silva, A.P. (2008). Nutritional quality of chestnut (*Castanea sativa* Mill.) cultivars from Portugal. *Food Chemistry*, 106, 976-984.

Boriani, M., Molinari, M. & Bazzoli, M. (2013). *Orthopelma mediator* (Thunberg) (Hymenoptera: Ichneumonidae) and the native parasitoid complex of *Dryocosmus kuriphilus* Yasumatsu (Hymenoptera: Cynipidae) in Lombardy (Italy). *Entomofauna Zeitschrift fur Entomologie*, 16, 201-204.

Borowiec, N., Thaon, M., Brancaccio, L., Warot, S., Vercken, E., Fauvergue, X., Ris, N. & Malausa, J.C. (2014). Classical biological control against the chestnut gall wasp *Dryocosmus kuriphilus* (Hymenoptera, Cynipidae) in France. *Plant Protection Quarterly*, 29, 7-10.

Bounous, G. (2005). The chestnut: a multipurpose resource for the new millennium. *Acta Horticulturae*, 693, 33-40.

Brussino, G., Bosio, G., Baudino, M., Giordano, R., Ramello, F. & Melika, G. (2002). Pericloso insetto esotico per il castagno europeo [Dangerous exotic insect for European chestnut]. *L'Informatore Agrario* 37, 59-61.

Bujdosó, B., Szabó, G., Bozsó, M. & Melika, G. (2013). Does the chestnut gall wasp, *Dryocosmus kuriphilus* (Hymenoptera: Cynipidae) already established in Hungary? *ISHS II European Congress on Chestnut 09-12, October*. Debrecen, Baia Mare, Modry Kamen.

Chang-Ti, T. & Yang, M. (2012). The occurrence and an unusual life cycle of *Dryocosmus kuriphilus* Yasumatsu (Hymenoptera: Cynipidae) in Taiwan. *ISHS II European Congress on Chestnut 09-12, October*. Debrecen, Baia Mare, Modry Kamen.

Colwell, R.K. (2009). EstimateS: Statistical estimation of species richness and shared species from samples. User's guide and application. Available at: http://purl.oclc.org/estimates, Version 8.2.

Cooper, W.R. & Rieske, L.K. (2007a). Community associates of an exotic gallmaker, *Dryocosmus kuriphilus* (Hymenoptera: Cynipidae), in Eastern North America. *Entomological Society of America*, 100, 236-244.

Cooper, W.R. & Rieske, L.K. (2007b). Review of the historic and current status of the Asian chestnut gall wasp in North America. *The Journal of the American Chestnut Foundation*, 21, 26-32.

Cortizo, E.V., Madriñan, M.L. & Madriñan, F.V. (1999). O castiñero: bioloxia e patoloxia [The chestnut: biology and pathology]. Santiago de Compostela: Consello da Cultura Galega.

EFSA (2010). Risk assessment of the oriental chestnut gall wasps Dryocosmus kuriphilus for the EU territory and identification and evaluation of risk management options. *EFSA Journal*, 114.

EPPO (2005). Data sheets on quarantine pests - *Dryocosmus kuriphilus*. *EPPO Bulletin*, 35, 422-424.

Everatt, M.J. (2015). Evaluation of the potential of control options used in Italy for the management of the oriental chestnut gall wasp, *Dryocosmus kuriphilus* Yasumatsu (Hymenoptera; Cynipidae), in the UK. Short Term Scientific Mission carried out as part of COST Action 1301 EuroCoppice, Department for Environment, Food and Rural Affairs.

Ferracini, C., Gonella, E., Ferrari, E., Saladini, M.A., Picciau, L., Tota, F., Pontini, M. & Alma, A. (2015). Novel insight in life cycle of *Torymus sinensis*, biocontrol agent of the chestnut gall wasp. *Biocontrol*, 60, 169-177.

Folcher, L., Bourguet, D., Thiery, D., Pelozuelo, L., Phalip, M., Weissenberd, A., Eychenne, N., Regnault, C.R. & Delos, M. (2011). Changes in parasitoid communities over time and space: a historical case study of maize pest *Ostrinia nubilialis*. *PLoS ONE*, 6, 253-274.

Francati, S., Alma, A., Ferracini, C., Pollini, A. & Dindo, M.L. (2015). Indigenous parasitoids associated with *Dryocosmus kuriphilus* in a chestnut prodution of Emília Romagna (Italy). *Bulletin of Insectology*, 68, 127-134.

Gauld, I. & Bolton, B. (1988). *The Hymenoptera*. London: Oxford University Press.

Gehring, E., Conedera, M. & Bosio, G. (2014). Edelkastaniengallwespe: Ein weiteres Problem für die Kastanie [Edelkastaniengallwespe: Another problem for the chestnut]. *Bündner Wald*, 67, 24-27.

Gibbs, M., Schonrogge, K., Alma, A., Melika, G., Quacchia, A., Stone, G.N. & Aebi, A. (2011). *Torymus sinensis*: a viable management option for the biological control of *Dryocosmus kuriphilus* in Europe? *BioControl*, 56, 527-538.

Gibson, G.A., Huber, J.T., & Woolley, J.B. (1997). Annotated keys to the genera of Nearctic Chalcidoidea (Hymenoptera). Ottawa: NRC-CNRC.

Gómez, J.F., Nieves, M.H., Torres, A.M., Askew, R.R. & Nieves-Aldrey, J.L. (2006). Los Chalcidoidea (Hymenoptera) associated with cynipid galls (Hymenoptera, Cynipidae) en la Comunidade de Madrid [The Chalcidoidea (Hymenoptera) com associados guts cinípideos (Hymenoptera, Cynipidae) in the Community of Madrid]. *Graellsia*, 62, 293-331.

Graziosi, I. & Rieske-Kinney, L.K. (2011). Interactions between an insect and a fungus pest: How American chestnut is impacted by two exotics. Proceedings of the 59th Annual Meeting of the Entomological Society of America.

Greathead, D.J. (1986). Parasitoids in classical biological control. In: J. Waage and D. Greathead (Eds.), *Insect Parasitoids* (pp. 289-318). New York: Academic Press.

Kos, K., Kriston, E. & Melika, G. (2015). Invasive chestnut gallwasp *Dryocosmus kuriphilus* (Hymenoptera: Cynipidae), its native parasitoid community and association with oak gall wasps in Slovenia. *European Journal of Entomology*, 112, 698-704.

Mansilla, J.P., Pérez, R., Pintos, C., Salinero, C. & Iglesias, C. (2000). Plagas y enfermedades del castaño en Galicia [Pests and diseases of the chestnut in Galicia]. Poio: Conselleria de Agricultura, Ganaderia e Política Agroalimentaria, Xunta de Galicia [Poio: Ministry of Agriculture, Livestock and Agricultural Policy, Government of Galicia].

Matošević, D. & Melika, G. (2013). Recruitment of native parasitoids to a new invasive host: first results of *Dryocosmus kuriphilus* parasitoid assemblage in Croatia. *Bulletin of Insectology*, 66, 231-238.

Matošević, D., Quacchia, A., Kriston, É. & Melika, G. (2014). Biological control of the invasive *Dryocosmus kuriphilus* (Hymenoptera: Cynipidae) - an overview & the first trials in Croatia. *South-East European Forestry*, 5, 3-12.

Matsuo, K., Yara, K., Kagoshima, K.M. & Moriya, S. (2011). Finding of *Torymus koreanus* (Hymenoptera: Cynipidae) in Japan. *Entomological Science*, 14, 100-102.

Monteiro, M.L. & Patrício, M.S. (2007). Conservação, regeneração e exploração do castanheiro [Conservation, regeneration and exploitation of the chestnut]. In: J. S. Silva (Ed.), *Do Castanheiro ao Teixo: As Outras Espécies Florestais* [From the Chestnut to the Yew: The Other Forest Species]. (pp. 51-78) Lisboa: Público, Fundação Luso-Americana para o Desenvolvimento.

Moriya, S., Inoue, K. & Mabuchi, M. (1989). The use of *Torymus sinensis* to control chestnut gall wasp, *Dryocosmus kuriphilus*, in Japan. *Technical Bulletin of Food and Fertilizer Technology Center*, 118, 1-12.

Moriya, S., Shiga, M. & Adachi, I. (2003). Classical biological control of the chestnut gall wasp in Japan. Proceedings of the 1st International Symposium on Biological Control of Arthropds. University of Hawaii. USDA-Foresty Service, Honolulu, Hawaii, 407-415.

Murakami, Y. (1981). The parasitoids of *Dryocosmus kuriphilus* Yasumatsu (Hymenoptera: Cynipidae) in Japan and the introduction of a promising natural enemy from China (Hymenopera: Chalcidoidea). *Journal of the Faculty of Agriculture, Kyushu University*, 25, 167-174.

Murakani, Y., Ao, H. & Chang., C. (1980). Natural enemies of the chestnut gall wasp in Hopei province, China (Hymenoptera: Chalcidoideia). *Applied Entomology and Zoology*, 15, 184-186.

Murakami, Y., Ohkubo, N., Moriya, S., Gyoutoku, Y., Kim, H.C. & Kim, K.J. (1995). Parasitoids of *Dryocosmus kuriphilus* (Hymenoptera: Cynipidae) in south Korea with particular reference to ecologically different types of *Torymus (syntomaspis) sinensis* (Hymenoptera: Torymidae). *Applied Entomology and Zoology*, 30, 277-284.

Murakami, Y., Toda, S., Gyoutoku, Y. & Mabuchi, M. (2001). Colonizations by imported *Torymus (syntomaspis) sinensis* Kamijo (Hymenoptera: Torymidae) of the chestnut gall wasp (Hymenoptera: Cynipidae). Success in the eighteenth year after release in Kumamoto. *Kyushu Plant Protection Research*, 47, 132-134.

Ôtake, A. (1980). Chestnut gall wasp *Dryocosmus kuriphilus* Yasumatsu (Hymenoptera: Cynipidae): a preliminary study on trend of adult emergence and some other ecological aspects related to the final stage of its life cycle. *Applied Entomology and Zoology*, 15, 96-105.

Ôtake, A. (1989). Chestnut gall wasp, *Dryocosmus kuriphilus* Yasumatsu (Hymenoptera: Cynipidae): analyses of records on cell contents inside galls and on emergence of wasps and parasitoids outside galls. *Applied Entomology and Zoology*, 24, 193-201.

Palmeri, V., Cascone, P., Campolo, O., Grande, S.B., Laudani, F., Malacrinò, A. & Guerrieri, E. (2014). Hymenoptera wasps associated with the Asian gall wasp of chestnut (*Dryocosmus kuriphilus*) in Calabria, Italy. *Phytoparasitica*, 42, 699-702.

Panzavolta, T., Bernardo, U., Bracalini, M., Cascone, P., Croci, F., Gebiola, M., Iodice, L., Tiberi, R. & Gerrieri, E. (2013). Native parasitoids associated with *Dryocosmus kuriphilus* in Tuscany, Italy. *Bulletin of Insectology*, 66, 195-201.

Perez, T. & Lopez, J.F. (2015). *Medidas Contra a Avespa Chinesa do Castiñeiro* (*Dryocosmus kuriphilus*) [Measures Against wasps Chinese chestnut (*Dryocosmus kuriphilus*)]. Santiago de Compostela: Xunta de Galicia, Consellería do Medio Rural o do Mar.

Quacchia, A., Moriya, S., Bosio, G., Scapin, G. & Alma, A. (2008). Rearing, release and settlement prospect in Italy of *Torymus sinensis*, the biological control agent of the chesnut gall wasp *Dryocosmus kuriphilus*. *BioControl*, 53, 829-839.

Quacchia, A., Piazza, E., Pavia, G. & Alma, A. (2011). Lotta biológica al cinipide del castagno – L'esperienza italiana con *Torymus sinensis* [Biological control of the chestnut gall wasp - The Italian experience with *Torymus sinensis*]. *Sherwood*, 177, 8-11.

Quacchia, A., Ferracini, C., Nicholls, J.A., Piazza, E., Saladini, M.A., Tota, F., Melika, G. & Alma, A. (2013). Chalcid parasitoid community associated with the invading pest *Dryocosmus kuriphilus* in north-western Italy. *Insect Conservation and Diversity*, 6, 114-123.

Quacchia, A., Moriya, S. & Bosio, G. (2014a). Effectiveness of *Torymus sinensis* in the biological control of *Dryocosmus kuriphilus* in Italy. Proceedings of the II European Congress on Chestnut, *Acta Horticulturae*, 1043, 199-204.

Quacchia, A., Morya, S., Askew, R. & Sconrogge, K. (2014b). *Torymus sinensis*: biology, host range and hybridization. Proceedings of the II European Congress on Chestnut, *Acta Horticulturae*, 1043, 105-112.

Sainti, F. & Maini, S. (2011). New association between *Dryocosmus kuriphilus* and *Torymus flavipes* in chestnut trees in the Bologna area (Italy): first results. *Bulletin of Insectology*, 64, 275-278.

Santor, C., Botta, R., Mellano, M.G., Beccaro, G.I, Bounous, G., Marinoni, D.T., Quacchia, A. & Alma, A. (2009) Evaluation of susceptibility of *Dryocosmus kuriphilus* Yasumatsu (Hymenoptera: Cypinidae) in *Castanea sativa* Miller and in hybrid cultivars. In: A. Soyhu and C. Mert, *Proceedings of the International Workshop on Chestnut Management in Mediterranean Countries, Acta Horticulturae*, 815, 289-298.

Santos, A. (2015). *Bioecologia da vespa-das-galhas-do-castanheiro,* Dryocosmus kuriphilus *Yasumatsu, no norte de Portugal*. Master Thesis. Bragança: Polytechnic Institute of Bragança.

Schonrogge, K., Moriya, S., Melika, G., Randle, Z., Begg, T., Aebi, A. & Stone, G.N. (2006). Early parasitoid recruitment in invading cynipid galls. In: K. Ozaki, J. Yukawa, T. Ohgushi and P.W. Price, *Galling Arthropods and Their Associates - Ecology and Evolution* (pp. 91-101) Tokyo: Springer-Verlag.

Speranza, S., Stacchiotti, M. & Paparatti, B. (2009). Endemic parasitoids of *Dryocosmus kuriphilus* Yasumatsu (Hymenoptera: Cynipidae) in Central Italy. In: Q. Ling and H. Hong-wen (Eds.) *Proceedings of the IV International Chestnut Symposium, Acta Horticulturae*, 844, 421-424.

Stone, G. N., Schoonrogge, K., Atkinson, R. J., Bellido, D. and Pujade-Villar, J. (2002). The population biology of oak gall wasps (Hymenoptera: Cynipidae). *Annual Review of Entomology*, 47, 633-668.

Viggiani, G. & Nugnes, F. (2010). Description of the larval stages of *Dryocosmus kuriphilus* Yasumatsu (Hymenoptera: Cynipidae), with notes on their phenology. *Journal of Entomological and Acarological Research*, 42, 39-45.

Yara, K., Sasawaki, T. & Kunimi, Y. (2010). Hybridization between introduced *Torymus sinensis* (Hymenoptera: Torymidae) and indigenous *T. beneficus* (late-spring strain), parasitoids of the Asian chestnut gall wasp *Dryocosmus kuriphilus* (Hymenoptera: Cynipidae). *Biological Control*, 54, 14-18.

Yasumatsu, K. (1951). A new *Dryocosmus* injurious to chestnut trees in Japan (Hym., Cynipidae). *Mushi*, 22, 89-93.

Zhi-Yong, Z. (2009). Study approaches on the chestnut gall wasp, *Dryocosmus kuriphilus* Yasumatsu in China. *Acta horticulturae*, 844, 425-432.

Biographical Sketches

Ana Lúcia Lobo Santos

School of Agriculture, Polytechnic Institute of Bragança,
Bragança, Portugal.

Education: Master in Agroecology

Research and Professional Experience: Scientific initiation scholarship at ESA – IPB between September of 2014 and June of 2016.

Professional Appointments: Trainer in POPH programs.

Honors: First student of Agroecology master.

Publications Last Three Years:

Santos, A. (2015). *Bioecologia da vespa-das-galhas-do-castanheiro,* Dryocosmus kuriphilus *Yasumatsu, no norte de Portugal.* Master Thesis in Agroecology. Bragança. Polytechnic Institute of Bragança.

Ambra Quacchia, PhD
GreenWood Service Srl

Professional Appointments:
Manager – Researcher GreenWood Service Srl

Education:
PhD in Agricultural, Forestry and Agro-food Sciences
University of Torino, Agriculture Faculty

Education at other institutions:
Institute of Evolutionary Biology, Ashworth Labs, Kings Buildings, University of
Edinburgh, United Kingdom, Scotland Stage from 07/2006 to 09/2006

Publications Last Three Years:

1. Quacchia A., Askew R., Moriya S. and Schönrogge K. (2014). *Torymus sinensis*: Biology, host range and hybridisation. ACTA HORTICULTURAE (ISHS) 1043:105-111.
2. Quacchia A., Moriya S. and Bosio G. (2014). Effectivenes of *Torymus sinensis* in the biological control of *Dryocosmus kuriphilus* in Italy. ACTA HORTICULTURAE (ISHS) 1043:199-204.
3. Matošević D., Quacchia A., Kriston É, Melika G. (2014). Biological Control of the Invasive *Dryocosmus kuriphilus* (Hymenoptera: Cynipidae) - an Overview and the First Trials in Croatia. SEEFOR 5 (1): 3-12
4. Quacchia A., Ferracini C., Nicholls J.A., Piazza E., Saladini M.A., Tota F., Melika G., Alma A., (2013). Chalcid parasitoid community associated with the invading pest *Dryocosmus kuriphilus* in north-western Italy. INSECT CONSERVATION AND DIVERSITY 6(2): 114-123.

INDEX

A

F

P

Q

R

S

T

U

V

W

Y